INTRODUCTION TO NUMBER THEORY

Essential Textbooks In Mathematics

ISSN: 2059-7657

The *Essential Textbooks in Mathematics* explores the most important topics in Mathematics that undergraduate students in Pure and Applied Mathematics are expected to be familiar with.

Written by senior academics as well lecturers recognised for their teaching skills, they offer in around 200 to 250 pages a precise, introductory approach to advanced mathematical theories and concepts in pure and applied subjects (e.g. Probability Theory, Statistics, Computational Methods, etc.).

Their lively style, focused scope and pedagogical material make them ideal learning tools at a very affordable price.

Published:

Introduction to Number Theory
by Richard Michael Hill (University College London, UK)

A Friendly Approach to Functional Analysis
by Amol Sasane (London School of Economics, UK)

A Sequential Introduction to Real Analysis
by J M Speight (University of Leeds, UK)

Essential Textbooks in Mathematics

INTRODUCTION TO NUMBER THEORY

Richard Michael Hill

University College London, UK

World Scientific

NEW JERSEY • LONDON • SINGAPORE • BEIJING • SHANGHAI • HONG KONG • TAIPEI • CHENNAI • TOKYO

Published by

World Scientific Publishing Europe Ltd.
57 Shelton Street, Covent Garden, London WC2H 9HE
Head office: 5 Toh Tuck Link, Singapore 596224
USA office: 27 Warren Street, Suite 401-402, Hackensack, NJ 07601

Library of Congress Cataloging-in-Publication Data
Names: Hill, Richard Michael, author.
Title: Introduction to number theory / by Richard Michael Hill (University College London, UK).
Other titles: Number theory
Description: New Jersey : World Scientific, 2018. | Series: Essential textbooks in mathematics | Textbook, with answers to some exercises. | Includes bibliographical references.
Identifiers: LCCN 2017044674| ISBN 9781786344717 (hc : alk. paper) | ISBN 9781786344892 (pbk : alk. paper)
Subjects: LCSH: Number theory--Textbooks.
Classification: LCC QA241 .H4845 2018 | DDC 512.7--dc23
LC record available at https://lccn.loc.gov/2017044674

British Library Cataloguing-in-Publication Data
A catalogue record for this book is available from the British Library.

For any available supplementary material, please visit
http://www.worldscientific.com/worldscibooks/10.1142/Q0137#t=suppl

Desk Editor: Herbert Moses/Jennifer Brough/Shi Ying Koe

Typeset by Stallion Press
Email: enquiries@stallionpress.com

Printed in Singapore

About the Author

Richard M. Hill grew up in England, and studied mathematics in Warwick and in Goettingen. He held research positions in the University of Goettingen and the Max Planck Institute in Bonn. Since 1994, he has worked in University College London. This book is based loosely around a course on number theory for second year undergraduates in UCL, which Richard has taught for several years.

Acknowledgments

Many thanks are due to my family for putting up with me while I wrote this book. I'd also like to thank my students for all the useful questions which they've asked me over the years.

Contents

Introduction

This book is intended as a textbook for a first course in number theory. In writing the book, rather than proving the most powerful general theorems, I've attempted to emphasize concrete questions about the integers. The reason for this emphasis is because most students enjoy solving such problems, and most students choose a first course in number theory because they hope it will be fun.

Some knowledge of university level mathematics is assumed. This is generally what one learns in the first year of studying for a mathematics degree in the UK. More precisely, I'll assume the reader is familiar with the following topics, in decreasing order of importance:

1. Standard methods of proof, such as proof by induction and proof by contradiction. The idea that an "if and only if" statement requires proof in both directions. Similarly, statements of the form "there exists a unique ..." require proof of "existence" and also of "uniqueness".
2. The notion of an equivalence relation.
3. The ideas of injective, surjective and bijective functions.
4. The reader should know what a group is, and know Lagrange's theorem, and what it means for two groups to be isomorphic.
5. The reader should know what a field is and be familiar with the fields $\mathbb{Q}$, $\mathbb{R}$ and $\mathbb{C}$.
6. In a few places in Chapters 4 and 5, it would help the reader to know a little real analysis. In particular, the reader should know what it means for a sequence to converge to a limit in $\mathbb{R}$.

Chapters 3–5 are based on a 30 hour lecture course for second year undergraduates at University College London. Chapters 1 and 2 and are more introductory in nature.

The book contains many exercises. Some of the exercises are simple applications of a method of calculation introduced in the text. Others need a little more thought. In a few cases, easy proofs are left as exercises in the main text; there are solutions to all the exercises at the end of the book. At the end of each chapter, there is a list of hints to some of the exercises.

Sage. Number theory is a very hands-on subject. One can write down examples to illustrate all the ideas in this book. The reader may find it helpful to be able to check calculations using a computer. For this, I recommend using the computer language `sage`, which was developed specifically for number theory. The language `sage` is free, and is available at http://www.sagemath.org/. At various places, I include examples of useful `sage` commands which are relevant to the theory being discussed. When I give examples of `sage` code, I'll use the following format:

```
1+1
```
    2

This will mean that if we type in `1+1`, then `sage` will return the answer 2. While I don't intend the book as a programming manual, I do hope to show how one can use `sage` as a calculator. A reader who wants to learn more about sage could look at the book [23].[a]

What's in the book. The book is written with the intention that the first three chapters are read consecutively and before Chapters 4 and 5. More experienced readers may find it more sensible to begin with Chapter 3. Chapters 4 and 5 are independent of each other, so they could be read in either order.

Chapter 1. The first chapter is on Euclid's algorithm for calculating highest common factors in the ring $\mathbb{Z}$. Other results are Bezout's Lemma, the Chinese Remainder Theorem and the uniqueness of factorization into primes.

Chapter 2. Chapter 2 is similar to Chapter 1, but with the ring $\mathbb{Z}$ replaced by the ring $\mathbb{F}[X]$ of polynomials over a field $\mathbb{F}$. It is shown that there is a version of Euclid's algorithm for this ring. As a consequence, it is shown that factorization of polynomials is unique. There are also

[a]The programming language used in this book is taken from Sage Math. All rights for this content are retained under the free, open-source GPL license.

methods described for checking whether a polynomial is irreducible, particularly over $\mathbb{Q}$. The concepts of unique factorization domain (UFD) and Euclidean domain are introduced; it's shown that $\mathbb{Z}[X]$ is a UFD, but not a Euclidean domain.

Chapter 3. Fermat's Little theorem and Euler's theorem are proved. Some properties of cyclotomic polynomials are discussed, and as a consequence, it is proved that $\mathbb{F}_p^\times$ is cyclic. Some applications to cryptography are described. The Quadratic Reciprocity Law is proved by the method of Gauss sums. This involves a brief discussion of rings of algebraic integers.

Chapter 4. p-Adic methods are introduced initially as methods for solving congruences modulo large powers of prime numbers. Two methods are described for solving such congruences. The first is Hensel's Lemma, and the second is the method of p-adic power series. The multiplicative group $(\mathbb{Z}/p^n)^\times$ is decomposed using p-adic logarithms and Teichmüller lifts. At the end of the chapter, we translate the methods and results into the language of p-adic integers and p-adic numbers.

Chapter 5. This chapter is a mixture of two topics. We study "quadratic rings", by which we shall mean the ring of algebraic integers in a quadratic field. We show that many of these rings have unique factorization. In the case of unique factorization, we also show how the decomposition of primes is related to the quadratic residue symbol. We describe the units in the real quadratic rings using continued fractions methods. The other topic of the chapter is Diophantine equations. We show how properties of quadratic rings may be used to solve certain Diophantine equations. We focus on norm equations such as $x^2 - dy^2 = a$ and also equations of the form $x^n = y^2 - d$. In cases where $\mathbb{Z}[\sqrt{d}]$ has unique factorization, we show how to solve such equations. At the end of the chapter, we solve such an equation with positive d by Skolem's p-adic method.

What's not in the book. We do not introduce the concept of an ideal in this book. As a consequence, rather than talking about principal ideal domains, we talk about UFDs. There is also no mention of any methods from complex analysis, so we do not give proofs of the Prime Number Theorem or Dirichlet's primes in arithmetic progressions theorem. We do not cover any Galois theory. We do not describe any sophisticated algorithms for primality testing or factorization.

Notation and conventions. The symbols $\mathbb{N}$, $\mathbb{Z}$, $\mathbb{Q}$, $\mathbb{R}$, $\mathbb{C}$ will mean the natural numbers, the integers, the rationals, the real numbers and the complex numbers, respectively. The symbol $\mathbb{F}$ will be used to mean a field. For a finite set S, we shall write $|S|$ for the *cardinality* of a set S, i.e. the number of elements. All rings will be commutative unital rings.

Chapter 1

Euclid's Algorithm

1.1 SOME EXAMPLES OF RINGS

Number theory began as the study of the natural numbers $\mathbb{N}$, the integers $\mathbb{Z}$ and the rational numbers $\mathbb{Q}$, and number theorists were concerned with questions about how integers factorize, which equations have solutions in integers, etc. The subject matter of number theory has shifted over the centuries. During the 19th century, several number theorists, beginning with Gauss, became interested in other sets of numbers, such as the *Gaussian integers*, which are defined as follows:

$$\mathbb{Z}[i] = \{x + iy : x, y \in \mathbb{Z}\}, \quad \text{where } i = \sqrt{-1}.$$

There were two reasons for this shift in interest. First, number theorists noticed that many of the theorems and methods which they had developed for solving problems with integers could also be used to solve similar problems with the Gaussian integers. Secondly, many problems concerning the integers have a solution which is most easily described using a bigger set of numbers. The most obvious example of this is the Quadratic Reciprocity Law which we shall see in a later chapter. This is a theorem about equations in $\mathbb{Z}$, but its most natural proof uses larger sets of numbers called the cyclotomic rings.

One thing which all these sets of numbers have in common is that they are all rings. In this book, the word "ring" will mean what is often called a "commutative unital ring", which is defined as follows:

Ring Axioms. A (commutative unital) *ring* is a set R with two binary operations $+$ and $\times$, satisfying the following axioms:

Addition Axiom. $(R, +)$ is an abelian group, with an identity element which we shall always call 0.

Multiplication Axiom. The operation $\times$ is associative and commutative, and there is a multiplicative identity element in R, which we shall always call 1. (We'll often write xy instead of $x \times y$).

Distributivity Axiom. For all elements $x, y, z \in R$, we have $(x+y)z = xz + yz$. In other words, we can expand out all brackets in the usual way.

We'll see several examples of rings in this book. Here are a few familiar examples.

The integers. We write $\mathbb{Z}$ for the ring of integers, i.e.

$$\mathbb{Z} = \{\ldots, -2, -1, 0, 1, 2, \ldots\}.$$

We have operations of addition and multiplication on $\mathbb{Z}$, and these operations obviously satisfy the Ring Axioms.

Fields. Every field is a ring, so in particular, the rational numbers $\mathbb{Q}$, the real numbers $\mathbb{R}$ and the complex numbers $\mathbb{C}$ are all rings. A field is a ring which satisfies two further axioms:

(1) The elements 0 and 1 in R are not equal.
(2) For every non-zero element $x \in R$, there is an element $x^{-1} \in R$, such that $xx^{-1} = 1$.

The integers modulo n. Let n be a positive integer. Two integers x and y are said to be *congruent modulo* n if $x - y$ is a multiple of n. This is written as

$$x \equiv y \mod n.$$

The relation of congruence modulo n is an equivalence relation. The *congruency class* $\bar{x}$ of an integer x is the set of integers congruent to x. This means that

$$\bar{x} = \{x + ny : y \in \mathbb{Z}\}.$$

For example, there are three congruency classes modulo 3, and they are

$$\bar{0} = \{\ldots, -6, -3, 0, 3, 6, \ldots\},$$
$$\bar{1} = \{\ldots, -5, -2, 1, 4, 7, \ldots\},$$
$$\bar{2} = \{\ldots, -4, -1, 2, 5, 8, \ldots\}.$$

Note that $\bar{3}$ is the same congruency class as $\bar{0}$ modulo 3. In fact, the congruence $x \equiv y \bmod n$ is equivalent to the equation $\bar{x} = \bar{y}$.

We'll write $\mathbb{Z}/n$ for the set of congruency classes modulo n. Since every integer is congruent to exactly one of the integers $0, 1, \ldots, n-1$, we have

$$\mathbb{Z}/n = \{\bar{0}, \bar{1}, \ldots, \overline{n-1}\}.$$

We can think of $\mathbb{Z}/n$ as the integers, with two integers regarded as equal when they are congruent modulo n.

Lemma 1.1. *Let x, x', y, y' be integers and let n be a positive integer. If $x \equiv x' \bmod n$ and $y \equiv y' \bmod n$, then $x + y \equiv x' + y' \bmod n$ and $xy \equiv x'y' \bmod n$.*

Proof. This is left as Exercise 1.1. (There is a solution at the end of the book.) □

Lemma 1.1 is quite useful for calculations. For example, suppose we'd like to know the congruency class of 26×37 modulo 10. According to the lemma, this is the same as the congruency class of 6×7 modulo 10 because $26 \equiv 6 \bmod 10$ and $37 \equiv 7 \bmod 10$.

The lemma is also used in defining operations of addition and multiplication on the set $\mathbb{Z}/n$, which allow us to regard $\mathbb{Z}/n$ as a ring. These operations are defined as follows:

$$\bar{x} + \bar{y} = \overline{(x + y)},$$
$$\bar{x} \times \bar{y} = \overline{(x \times y)}.$$

For example, in $\mathbb{Z}/8$, we have

$$\overline{33} + \overline{12} = \overline{45} = \bar{5} \quad \text{and} \quad \overline{33} \times \overline{12} = \overline{396} = \bar{4}.$$

One needs to be a little careful with definitions like this. For example, in $\mathbb{Z}/8$, we have $\overline{33} = \overline{17}$ and $\overline{12} = \overline{20}$. Therefore, we require $\overline{33} \times \overline{12}$ to be the same congruency class as $\overline{17} \times \overline{20}$. However, Lemma 1.1 tells us that this is the case. Indeed, in the language of congruency classes, the lemma says that if $\overline{x} = \overline{x'}$ and $\overline{y} = \overline{y'}$, then $\overline{x+y} = \overline{x'+y'}$ and $\overline{xy} = \overline{x'y'}$.

To show that $\mathbb{Z}/n$ is a ring (with the operations $+$ and $\times$ just defined), one needs to check the Ring Axioms. As an example, we'll show that the operation of multiplication on $\mathbb{Z}/n$ defined above is associative. For this, we need to check that for all elements $\bar{x}, \bar{y}, \bar{z} \in \mathbb{Z}/n$, we have

$$(\bar{x} \times \bar{y}) \times \bar{z} = \bar{x} \times (\bar{y} \times \bar{z}).$$

After unwinding the definition of multiplication, this amounts to proving the congruence

$$(xy)z \equiv x(yz) \mod n.$$

However, the congruence is obviously true: the two sides are actually equal as integers, so they are certainly congruent modulo n. In fact, each ring axiom for $\mathbb{Z}/n$ can be expressed as a congruence. This congruence is implied by the corresponding ring axiom for $\mathbb{Z}$, which is the same statement with the congruence relation replaced by "=". In this way, we see that the ring axioms for $\mathbb{Z}/n$ are consequences of the ring axioms for $\mathbb{Z}$.

In practice, the notation $\bar{x}$ for a congruency class is very inconvenient. It does not tell us the value of n, and it's a lot of extra work writing a bar on top of every symbol. Instead, we write the elements of $\mathbb{Z}/n$ as integers, but replace the "=" sign with congruence modulo n. In particular, we'll often write $\mathbb{Z}/n = \{0, 1, \ldots, n-1\}$, when we really mean that the elements of $\mathbb{Z}/n$ are congruency classes.

Example. The elements of $\mathbb{Z}/5$ are the congruency classes of 0, 1, 2, 3 and 4. Addition and multiplication are given by the following tables:

$+$	0	1	2	3	4
0	0	1	2	3	4
1	1	2	3	4	0
2	2	3	4	0	1
3	3	4	0	1	2
4	4	0	1	2	3

$\times$	0	1	2	3	4
0	0	0	0	0	0
1	0	1	2	3	4
2	0	2	4	1	3
3	0	3	1	4	2
4	0	4	3	2	1

One can see from the multiplication table that $\mathbb{Z}/5$ is actually a field. This is because $1 \not\equiv 0 \bmod 5$, and every element apart from 0 has an inverse for multiplication:

$$1 \times 1 \equiv 1 \bmod 5, \quad 2 \times 3 \equiv 1 \bmod 5, \quad 4 \times 4 \equiv 1 \bmod 5.$$

We'll see in Theorem 1.2 that whenever n is a prime number, the ring $\mathbb{Z}/n$ is a field.

Exercise 1.1. Show that if $x \equiv x' \bmod n$ and $y \equiv y' \bmod n$, then $x + y \equiv x' + y' \bmod n$ and $xy \equiv x'y' \bmod n$ (i.e. prove Lemma 1.1).

1.2 EUCLID'S ALGORITHM

Let's suppose that d and m are integers. We say that d is a *factor* of m, or d *divides* m, or m is a *multiple* of d if $m = dx$ for some integer x. This is written as

$$d|m.$$

If d is a factor of two integers n and m, then we call d a *common factor* of n and m. Unless n and m are both zero, they have only finitely many common factors. There is therefore the *highest common factor* which is written $\mathrm{hcf}(n, m)$ or sometimes $\gcd(n, m)$. We'll say that n and m are *coprime* if their only common factors are 1 and -1, or equivalently if their highest common factor is 1.

Suppose that n and m are integers with $n \geq m > 0$. Dividing n by m with remainder, we may find integers q and r, such that

$$n = qm + r, \quad 0 \leq r < m. \tag{1.1}$$

The integers q and r are called the *quotient* and the *remainder*. All we're saying here is that the remainder on dividing an integer by m will always be less than m. From Equation (1.1), we can see that

Every common factor of m and r is also a factor of n.

Rearranging Equation (1.1), we have $r = n - qm$, and from this, we can see that

Every common factor of n and m is also a factor of r.

The last two statements show that the common factors of n and m are exactly the same as the common factors of m and r. In particular, this implies

$$\mathrm{hcf}(n, m) = \mathrm{hcf}(m, r). \tag{1.2}$$

As the new numbers m and r are smaller than the original numbers n and m, we have reduced the problem of calculating $\mathrm{hcf}(n, m)$ to the smaller problem of calculating $\mathrm{hcf}(m, r)$. Indeed, as long as r is non-zero, we can reduce the size of the problem further by dividing m by r with remainder. Here is an example of how this method works.

Example. We'll calculate $\text{hcf}(106, 42)$. The first step is to divide 106 by 42 with remainder:

$$106 = 2 \times 42 + 22.$$

Equation (1.2) shows that $\text{hcf}(106, 42) = \text{hcf}(42, 22)$. To calculate $\text{hcf}(42, 22)$, we again divide with remainder:

$$42 = 1 \times 22 + 20.$$

This shows that $\text{hcf}(42, 22) = \text{hcf}(22, 20)$. Continuing in this way, we calculate as follows:

$$22 = 1 \times 20 + 2,$$

$$20 = 10 \times 2 + 0.$$

Therefore, $\text{hcf}(22, 20) = \text{hcf}(20, 2) = \text{hcf}(2, 0)$. Since every integer is a factor of 0, the highest common factor of 2 and 0 is just the largest factor of 2, which is 2. Putting all this together, we've shown that $\text{hcf}(106, 42) = \text{hcf}(2, 0) = 2$.

The algorithm used above for calculating $\text{hcf}(106, 42)$ is called *Euclid's algorithm*. Here is a more formal description of the algorithm.

Euclid's Algorithm.

(1) We begin with two integers n, m and let's assume that $n \geq m \geq 0$.
(2) Define a finite sequence of integers

$$r_1 \geq r_2 > \cdots > r_N > r_{N+1} = 0$$

recursively as follows: $r_1 = n$, $r_2 = m$, and if $r_s \neq 0$, then r_{s+1} is the remainder on dividing r_{s-1} by r_s.
(3) The highest common factor of n and m is the number r_N, i.e. the last non-zero remainder. To see why this is the case, we use Equation (1.2) several times:

$$\text{hcf}(n, m) = \text{hcf}(r_1, r_2) = \text{hcf}(r_2, r_3) = \cdots = \text{hcf}(r_N, 0) = r_N.$$

Apart from being useful for calculating the highest common factors, Euclid's algorithm also has significant theoretical consequences. This is because it allows us to prove results concerning highest common factors by induction; the inductive step uses Equation (1.2). One of the most important examples of this kind of proof is the following lemma.

Lemma 1.2 (Bezout's Lemma). *Let n and m be integers which are not both 0. Then there exist integers h and k such that*

$$\text{hcf}(n,m) = hn + km. \tag{1.3}$$

Before proving Bezout's Lemma, we'll describe a method which actually finds integers h and k which solve Equation (1.3). This method is just an extension of Euclid's algorithm. Suppose we divide n by m with remainder:

$$n = qm + r, \quad 0 \leq r < m.$$

The numbers m, r are smaller than n, m, and so the problem of solving the equation $\text{hcf}(m,r) = h'm + k'r$ seems easier than our original problem. Indeed if the remainder r is zero, then there is a simple solution $h' = 1$, $k' = 0$. Let's suppose we already know an expression $\text{hcf}(m,r) = h'm + k'r$, then since $\text{hcf}(n,m) = \text{hcf}(m,r)$, we have

$$\text{hcf}(n,m) = h'm + k'(n - qm) = hn + km,$$

where $h = k'$ and $k = h' - k'q$.

Example. We'll find integers h, k such that $\text{hcf}(106, 42) = 102h + 42k$. Recall that we calculated $\text{hcf}(106, 42)$ as follows:

$$\begin{aligned} \mathbf{106} &= 2 \times \mathbf{42} + \mathbf{22}, \\ \mathbf{42} &= 1 \times \mathbf{22} + \mathbf{20}, \\ \mathbf{22} &= 1 \times \mathbf{20} + \mathbf{2}, \\ \mathbf{20} &= 10 \times \mathbf{2} + \mathbf{0}. \end{aligned}$$

The numbers in bold are the sequence r_n in Euclid's algorithm. The highest common factor is the last non-zero remainder, which in this case is 2. Using the third line in the calculation above, we can write 2 as a multiple of 22 plus a multiple of 20:

$$\mathbf{2} = \mathbf{22} - 1 \times \mathbf{20}.$$

Using the second line, we can replace the 20 in this equation by an expression in terms of 42 and 22:

$$\mathbf{2} = \mathbf{22} - 1 \times (\mathbf{42} - \mathbf{22}).$$

Expanding out the bracket and collecting terms together, we get an expression for 2 in terms of 22 and 42:

$$\mathbf{2} = 2 \times \mathbf{22} - \mathbf{42}.$$

Using the top line of the calculation, we can replace the 22 by an expression in terms of 106 and 42:

$$\mathbf{2} = 2 \times (\mathbf{106} - 2 \times \mathbf{42}) - \mathbf{42}.$$

As before, we can simplify this:

$$\mathbf{2} = 2 \times \mathbf{106} - 5 \times \mathbf{42}.$$

We've shown that $h = 2$, $k = -5$ is a solution to Equation (1.3). Of course, this is not the only solution; if we add 42 to h and subtract 106 from k, then we will obtain another solution, so in fact there are always infinitely many solutions to Equation (1.3).

Proof of Bezout's Lemma. We can replace n and m by $-n$ and $-m$ without altering their highest common factor, and so we may assume without loss of generality that n and m are both greater than or equal to 0. Also, we'll assume that $m \leq n$. We'll prove the statement by induction on m. If $m = 0$, then $\text{hcf}(n, m) = n$, and the lemma is true with $h = 1$ and $k = 0$. Assume now that $m > 0$ and that the lemma is true for all pairs of integers n', m' with $0 \leq m' < m$. As $m > 0$, we may divide n by m with remainder:

$$n = qm + r, \quad 0 \leq r < m.$$

As we saw in Equation (1.2), $\text{hcf}(n, m) = \text{hcf}(m, r)$. However, since $r < m$, the inductive hypothesis implies that there exist integers h' and k' such that

$$\text{hcf}(m, r) = h'm + k'r.$$

Rearranging this, we get

$$\text{hcf}(n, m) = h'm + k'(n - qm) = hn + km,$$

where $h = k'$ and $k = h' - k'q$. □

Exercise 1.2. Find integers h and k such that $123h + 136k = 1$.

One can answer questions like Exercise 1.2 using `sage`. For example, the command `gcd(1349,6035)` will give the highest common factor of 1349 and 6035. To find integers h and k satisfying Equation (1.3), we use the command `xgcd(1349,6035)`. This will return the highest common factor and the integers h and k. For example,

```
gcd(1349,6035)
```

71

```
xgcd(1349,6035)
```

(71, 9, −2)

```
9 * 1349 - 2 * 6035
```

71

Exercise 1.3. Let n and m be integers which are not both zero. Show that an integer a can be written in the form $rn + sm$ with $r, s \in \mathbb{Z}$ if and only if a is a multiple of $\mathrm{hcf}(n, m)$.

Exercise 1.4. Show that d is a common factor of n and m if and only if d is a factor of $\mathrm{hcf}(n, m)$.

Exercise 1.5. Let n and m be positive integers. We shall write $\mathrm{lcm}(n, m)$ for the smallest positive integer which is a multiple of both n and m. Show that

$$\mathrm{hcf}(n, m) \cdot \mathrm{lcm}(n, m) = nm.$$

1.3 INVERTIBLE ELEMENTS MODULO n

Let's suppose that R is a ring. An element $x \in R$ is called *invertible*, or a *unit* if there is an element y in R, such that $xy = 1$. The element y is called the *inverse* of x. Note that x cannot have more than one inverse, since if z is also an inverse then by the Ring Axioms, we have

$$y = y \cdot 1 = y(xz) = (yx)z = 1 \cdot z = z.$$

It therefore makes sense to write x^{-1} for the inverse of x, when this exists.

We'll write $R^\times$ for the set of invertible elements in a ring R. For example, in the ring $\mathbb{Z}$, only the elements 1 and -1 are invertible, so $\mathbb{Z}^\times = \{1, -1\}$. In contrast, if $\mathbb{F}$ is a field, then all the non-zero elements have inverses, so $\mathbb{F}^\times$ is the set of all non-zero elements of $\mathbb{F}$.

Proposition 1.1. *For any ring R, the set $R^\times$ is a group with the operation of multiplication.*

Proof. If x and y are both invertible, then xy is invertible with inverse $x^{-1}y^{-1}$. Therefore, $R^\times$ is closed under multiplication. We must check the three group axioms.

Associativity. The operation of multiplication is certainly associative on $R^\times$ because associativity is one of the Ring Axioms.

Identity. The element $1 \in R$ is invertible with inverse 1, so $R^\times$ has an identity element.

Inverses. If x is invertible, then x^{-1} is also invertible (with inverse x). Therefore, every element of $R^\times$ has an inverse in $R^\times$. □

We'll now consider which elements of the ring $\mathbb{Z}/n$ are invertible. In other words, given an element $x \in \mathbb{Z}/n$, we want a way of finding out whether there is a $y \in \mathbb{Z}/n$ such that $xy \equiv 1 \bmod n$. To illustrate this problem, we look at an example. The ring $\mathbb{Z}/10$ has elements $\{0, 1, 2, 3, 4, 5, 6, 7, 8, 9\}$, and multiplication in this ring is given by the following table:

$\times$	0	1	2	3	4	5	6	7	8	9
0	0	0	0	0	0	0	0	0	0	0
1	0	①	2	3	4	5	6	7	8	9
2	0	2	4	6	8	0	2	4	6	8
3	0	3	6	9	2	5	8	①	4	7
4	0	4	8	2	6	0	4	8	2	6
5	0	5	0	5	0	5	0	5	0	5
6	0	6	2	8	4	0	6	2	8	5
7	0	7	4	①	8	5	2	9	6	3
8	0	8	6	4	2	0	8	6	4	2
9	0	9	8	7	6	5	4	3	2	①

To find out whether an element x has an inverse, we look along row x to see if any multiple of x is 1 (in the table, the 1's are circled). From the table, we can see that only 1, 3, 7 and 9 have inverses, so $(\mathbb{Z}/10)^\times = \{1, 3, 7, 9\}$.

It's quite time consuming to write out the whole table, just to find out whether an element has an inverse. The following proposition is a much quicker way of checking.

Proposition 1.2. *Let n be a positive integer and let $x \in \mathbb{Z}/n$. Then x is invertible modulo n if and only if x and n are coprime.*

Proof. Suppose first that x is invertible, and that y is an inverse of x modulo n. This means that $1 = xy + nz$ for some integer z. It follows that every common factor of x and n must also be a factor of 1. This implies $\text{hcf}(x, n) = 1$, so x and n are coprime.

Conversely, suppose that $\text{hcf}(x, n) = 1$. By Bezout's Lemma, there exist integers h and k such that $hx + kn = 1$. This implies $hx \equiv 1 \bmod n$, so h is the inverse of x. □

Looking at the proof of Proposition 1.2, we see that in order to find the inverse of an element x in $\mathbb{Z}/n$, it's sufficient to find a solution to $1 = hx + kn$ by the method described just after Bezout's Lemma. Reducing this equation modulo n, we see that h is the inverse of x.

Example. We'll find the inverse of 29 in $\mathbb{Z}/100$, or equivalently we'll find an integer x such that $29x \equiv 1 \bmod 100$.

In order for such an inverse to exist, the numbers 100 and 29 must be coprime. To check this, we go through Euclid's algorithm with 100 and 29. As before, the sequence r_n is in bold.

$$\begin{aligned}
\mathbf{100} &= 3 \times \mathbf{29} + \mathbf{13}, \\
\mathbf{29} &= 2 \times \mathbf{13} + \mathbf{3}, \\
\mathbf{13} &= 4 \times \mathbf{3} + \mathbf{1}, \\
\mathbf{3} &= 3 \times \mathbf{1} + \mathbf{0}.
\end{aligned}$$

The highest common factor is the last non-zero remainder, which in this case is 1. Hence, 100 and 29 are coprime, so 29 is invertible modulo 100. To find $29^{-1} \bmod 100$, we find a solution to $1 = 100h + 29k$ as before.

$$\begin{aligned}
\mathbf{1} &= \mathbf{13} - 4 \times \mathbf{3} \\
&= \mathbf{13} - 4 \times (\mathbf{29} - 2 \times \mathbf{13}) \\
&= 9 \times \mathbf{13} - 4 \times \mathbf{29} \\
&= 9 \times (\mathbf{100} - 3 \times \mathbf{29}) - 4 \times \mathbf{29} \\
&= 9 \times \mathbf{100} - 31 \times \mathbf{29}.
\end{aligned}$$

Reducing the final line of this calculation modulo 100, we find that

$$1 \equiv -31 \times 29 \mod 100.$$

Therefore, the inverse of 29 modulo 100 is -31. We might sometimes want to express the inverse as an integer between 0 and 99. To change -31 into this form, we simply add 100 to it, which will not change the congruency class modulo 100. We therefore have

$$29^{-1} \equiv 69 \mod 100.$$

Looking at this example, we can see that our method of finding 29^{-1} is rather fast, when compared with the most naive method where one simply multiplies all the numbers from 1 to 99 by 29 until one finds a solution to $29x \equiv 1 \bmod 100$.

Exercise 1.6. Find all elements of $(\mathbb{Z}/12)^{\times}$ and write down a multiplication table for this group.

To answer a question like Exercise 1.6 in `sage`, we first define `R` to be the ring $\mathbb{Z}/n$ using the command `IntegerModRing`

```
R=IntegerModRing(8)
R
```

$\mathbf{Z}/8\mathbf{Z}$

To list the units in $\mathbb{Z}/8$, we use the following command:

```
R.list_of_elements_of_multiplicative_group()
```

[1, 3, 5, 7]

This shows that $(\mathbb{Z}/8)^{\times} = \{1, 3, 5, 7\}$. We can produce a multiplication table of just these elements as follows:

```
G=R.list_of_elements_of_multiplicative_group()
R.multiplication_table(names='elements',elements=G)
```

*	1	3	5	7
1	1	3	5	7
3	3	1	7	5
5	5	7	1	3
7	7	5	3	1

Exercise 1.7. Using Euclid's algorithm, find the inverses of 9 and 23 in $\mathbb{Z}/175$.

There are several ways to calculate inverses in `sage`. We'll illustrate these methods by calculating 43^{-1} modulo 120. The easiest way is to simply reduce modulo 120 using the operation `%`. If `a` and `b` are integers, then `a%b` will return the remainder on dividing `a` by `b`. The operation also works when `a` is a rational number, as long as its denominator is invertible modulo `b`.

For example,

```
1/43 % 120
```

67

```
43^-1 % 120
```

67

There is also a command `inverse_mod`, which does the same thing.

```
inverse_mod(43,120)
```

67

One can also use the command `mod(43,120)` to define the congruency class of 43 modulo 120 and then find the inverse of this congruency class.

```
mod(43,120)^-1
```

67

We can also define `R` to be the ring $\mathbb{Z}/120$ and then use `R(43)` to indicate the element 43 in the ring $\mathbb{Z}/120$.

```
R=IntegerModRing(120)
R(43)^-1
```

67

The third and fourth methods are subtly different from the first and second methods. This is because the third and fourth methods output the congruency class of 67 modulo 120, whereas the first two methods output the integer 67. This difference is apparent if we use the answer in a further calculation:

```
2*inverse_mod(43,120)
```

134

```
2*R(43)^-1
```

14

1.4 SOLVING LINEAR CONGRUENCES

Suppose we'd like to solve a congruence of the form

$$ax \equiv b \mod n, \tag{1.4}$$

where we are given a, b and n and we would like to find all the integer solutions x. Assume first that a and n are coprime. By Proposition 1.2, this implies a is invertible modulo n. Multiplying both sides of Equation (1.4) by a^{-1}, we have the solution:

$$x \equiv a^{-1}b \mod n.$$

Example. We'll solve the congruence

$$5x \equiv 11 \mod 13.$$

We go through Euclid's algorithm starting with 13 and 5. As before, we write the sequence r_n from Euclid's Algorithm in bold.

$$\mathbf{13} = 2 \times \mathbf{5} + \mathbf{3},$$
$$\mathbf{5} = 1 \times \mathbf{3} + \mathbf{2},$$
$$\mathbf{3} = 1 \times \mathbf{2} + \mathbf{1}.$$

The next remainder would be 0 because our next step would be to divide by 1. Therefore, the highest common factor of 13 and 5 is 1. This shows us that 5 is invertible modulo 13, and we continue with our method for finding the inverse. The next step is to find an expression of the form $1 = h \cdot 13 + k \cdot 5$, working backwards through the calculation we've just done.

$$\begin{aligned} \mathbf{1} &= \mathbf{3} - \mathbf{2} \\ &= \mathbf{3} - (\mathbf{5} - \mathbf{3}) \\ &= 2 \times \mathbf{3} - \mathbf{5} \\ &= 2(\mathbf{13} - 2 \times \mathbf{5}) - \mathbf{5} \\ &= 2 \times \mathbf{13} - 5 \times \mathbf{5}. \end{aligned}$$

Hence, $-5 \times 5 \equiv 1 \bmod 13$ so $5^{-1} \equiv -5 \bmod 13$. Now, we can solve the congruence:

$$x \equiv -5 \times 11 = -55 \equiv 10 \mod 13.$$

We can check that this is indeed a solution: $5 \times 10 = 50 \equiv 11 \bmod 13$.

This method will only work if a is coprime to n, since otherwise a would have no inverse. However, we can still solve the congruence in Equation (1.4) even when a and n have a common factor. To do this, we use the following two exercises.

Exercise 1.8. Show that if a is a factor of n but is not a factor if b, then the congruence $ax \equiv b \bmod n$ has no solutions $x \in \mathbb{Z}$.

Exercise 1.9. Show that if a is a common factor of b and n, then the congruence $ax \equiv b \bmod n$ is equivalent to $x \equiv \frac{b}{a} \bmod \frac{n}{a}$.

The exercises give us a general method for solving Equation (1.4):

Case 1. Suppose a is invertible module n, i.e. there is an element $a^{-1} \in \mathbb{Z}/n$ such that $a \times a^{-1} \equiv 1 \bmod n$. Then we can multiply both sides of the congruence by a^{-1} to get

$$x \equiv a^{-1}b \mod n.$$

Case 2. Suppose that n is a multiple of a. In particular, n and a are not coprime, so a has no inverse modulo n. If b is not a multiple of a, then the congruence has no solutions. If on the other hand b is a multiple of a, then the solution is

$$x \equiv \frac{b}{a} \mod \frac{n}{a}.$$

(Note that both $\frac{b}{a}$ and $\frac{n}{a}$ are integers.)

Example.

$$7x \equiv 84 \mod 490.$$

Note that 7 is a common factor of 84 and 490, so the solution is $x \equiv 12 \bmod 70$.

Example.

$$7x \equiv 85 \mod 490.$$

Note that 7 is a factor of 490, but is not a factor of 85, so this congruence has no solutions.

Example.

$$6x \equiv 3 \mod 21.$$

This is in neither case 1 nor case 2, since 6 is neither coprime to 21 nor a factor of 21. However, we can rewrite the congruence as

$$3(2x) \equiv 3 \mod 21.$$

Using case 2, we deduce that

$$2x \equiv 1 \mod 7.$$

Now 2 and 7 are coprime, so we use case 1 to get

$$x \equiv 2^{-1} \equiv 4 \mod 7.$$

Exercise 1.10. Solve each of the following congruences:

$$4x \equiv 3 \mod 71, \qquad 6x \equiv 12 \mod 71,$$

$$12x \equiv 20 \mod 72, \qquad 20x \equiv 12 \mod 72,$$

$$55x \equiv 44 \mod 121.$$

One can solve such congruences in `sage` using the command `solve_mod`. For example, to solve the first congruence in the exercise above, we use the command

```
solve_mod(4*x==3, 71)
```

[(54)]

1.5 THE CHINESE REMAINDER THEOREM

Suppose that x is an integer which is congruent to 7 modulo 10. The congruence $x \equiv 7 \bmod 10$ means that $x - 7$ is a multiple of 10. Obviously, this implies that $x - 7$ is a multiple of 5 and is also a multiple of 2. Hence, the congruence $x \equiv 7 \bmod 10$ implies both $x \equiv 7 \bmod 5$ and $x \equiv 7 \bmod 2$. However, we know nothing of the congruency class of x modulo 3. For example, x could be the integer 7, which is congruent to 1 modulo 3, or x could be the integer 17 which is congruent to 2 modulo 3, or x could be the integer 27 which is congruent to 0 modulo 3. The difference here is that 2 and 5 are factors of 10, whereas 3 is coprime to 10. In general, if $x \equiv y \bmod nm$, then $x \equiv y \bmod n$ and $x \equiv y \bmod m$.

One could ask whether the converse to this is true: if we know the congruency classes of an integer x modulo n and modulo m, then can we recover its congruency class modulo nm. This question is answered by the next result.

Theorem 1.1 (Chinese Remainder Theorem). *Assume n and m are coprime and let $a \in \mathbb{Z}/n$, $b \in \mathbb{Z}/m$. Then there is a unique $x \in \mathbb{Z}/nm$ such that*

$$x \equiv a \bmod n, \quad x \equiv b \bmod m.$$

Before proving the theorem, we'll discuss the general strategy for proving this kind of result. The theorem states that there is a *unique* solution x in $\mathbb{Z}/nm$, and so the proof is in two steps. First, we show that there is at least one solution; this is called the *existence* part of the proof.

After that, we shall show that any two solutions are actually congruent modulo nm; this is called the *uniqueness* part of the proof.

There is another way of thinking about these two steps, which we shall use here. Recall that for two sets X and Y, the *Cartesian product* $X \times Y$ is the set of pairs (x, y), where x is an element of X and y is an element of Y. For example,

$$\{1, 3, 4\} \times \{2, 3\} = \{(1, 2), (1, 3), (3, 2), (3, 3), (4, 2), (4, 3)\}.$$

We have a function $\mathcal{C}$ from $\mathbb{Z}/nm$ to the Cartesian product $\mathbb{Z}/n \times \mathbb{Z}/m$, which takes the congruency class of x modulo nm to the pair ($x \bmod n, x \bmod m$). The existence part of the proof is showing that $\mathcal{C}$ is a surjective function, since it shows that for any a and b, there is an integer x such that $\mathcal{C}(x) = (a, b)$. The uniqueness part of the proof is showing that $\mathcal{C}$ is injective.

Proof. (*Existence*) Since n and m are coprime, we can find integers h and k such that $hn + km = 1$. Let $x = hnb + kma$. We'll show that x is a solution to both congruences. From the definition of x, we have

$$x \equiv kma \mod n.$$

However, the equation $hn + km = 1$ implies $km \equiv 1 \bmod n$. Therefore,

$$x \equiv a \mod n.$$

A similar argument shows that $x \equiv b \bmod m$.

(*Uniqueness*) By proving the existence part of the theorem, we've shown that the map $\mathcal{C} : \mathbb{Z}/nm \to \mathbb{Z}/n \times \mathbb{Z}/m$ given by

$$\mathcal{C}(x) = (x \bmod n, x \bmod m)$$

is surjective. To prove the uniqueness part of the theorem, we must show that this map is also injective. However, both sets $\mathbb{Z}/nm$ and $\mathbb{Z}/n \times \mathbb{Z}/m$ have exactly nm elements, so any surjective map from one set to the other is also injective. □

If we look again at the existence part of this proof, then we see that it actually gives us a formula $x = hnb + kma$ for the solution to the simultaneous congruences. This allows us to solve these congruences easily in practice, which is rather more than just knowing that there is a solution.

Example. We'll solve the simultaneous congruences

$$x \equiv 3 \bmod 8, \quad x \equiv 4 \bmod 5.$$

The numbers 5 and 8 are coprime, so there will be integers h and k satisfying $8h + 5k = 1$. The solution to the simultaneous congruences is given by

$$x \equiv h \cdot 8 \cdot 4 + k \cdot 5 \cdot 3 \mod 5 \times 8.$$

We can find the integers h and k as before by Euclid's algorithm.

$$\begin{array}{ll} \mathbf{8} = \mathbf{5} + \mathbf{3} & 1 = \mathbf{3} - \mathbf{2} \\ \mathbf{5} = \mathbf{3} + \mathbf{2} & \quad = \mathbf{3} - (\mathbf{5} - \mathbf{3}) \\ \mathbf{3} = \mathbf{2} + \mathbf{1} & \quad = 2 \times \mathbf{3} - \mathbf{5} \\ & \quad = 2 \times (\mathbf{8} - \mathbf{5}) - \mathbf{5} \\ & \quad = 2 \times \mathbf{8} - 3 \times \mathbf{5}. \end{array}$$

This shows that we may take $h = 2$ and $k = -3$. Hence, the solution is

$$x \equiv 2 \cdot 8 \cdot 4 - 3 \cdot 5 \cdot 3 \equiv 19 \mod 40.$$

We can easily check that $x = 19$ is a solution and we have made no mistake, since $19 \equiv 3 \bmod 8$ and $19 \equiv 4 \bmod 5$. The uniqueness part of the Chinese Remainder Theorem tells us that there are no other solutions in $\mathbb{Z}/40$.

Remark. The Chinese Remainder Theorem gives us a method for solving congruences modulo nm by simply solving modulo n and modulo m. For example, suppose we would like to solve

$$x^2 \equiv 2 \mod 119.$$

We can factorize $119 = 7 \times 17$, so it's really enough to find solutions modulo 7 and modulo 17 and then put the solutions together using the Chinese Remainder Theorem. The numbers 7 and 17 are small enough that we can find the solutions by hand:

$$x \equiv \pm 3 \bmod 7, \quad x \equiv \pm 6 \bmod 17.$$

To put the solutions together, we go through Euclid's algorithm with 7 and 17:

$$\begin{array}{ll} \mathbf{17} = 2 \times \mathbf{7} + \mathbf{3} & \mathbf{1} = \mathbf{7} - 2 \times \mathbf{3} \\ \mathbf{7} \;\, = 2 \times \mathbf{3} + \mathbf{1} & \quad = \mathbf{7} - 2(\mathbf{17} - 2 \times \mathbf{7}) \\ & \quad = 5 \times \mathbf{7} - 2 \times \mathbf{17}. \end{array}$$

Hence, the solutions modulo 119 are

$$\begin{aligned} x &\equiv 5 \cdot 7 \cdot (\pm 6) - 2 \cdot 17 \cdot (\pm 3) \\ &\equiv \pm 28 \pm 17 \mod 119. \end{aligned}$$

The two $\pm$ signs may be chosen independently of each other, and so there are four solutions modulo 119, which are given by

$$x \equiv 45, \ -45, \ 11 \text{ or } -11 \mod 119.$$

Exercise 1.11. Solve each of the following pairs of simultaneous congruences:

$$\text{(a) } x \equiv 3 \bmod 8, \qquad x \equiv 4 \bmod 5,$$
$$\text{(b) } x \equiv 2 \bmod 25, \qquad x \equiv 1 \bmod 19.$$

Exercise 1.12. Prove that the simultaneous congruences

$$x \equiv a \bmod n, \quad x \equiv b \bmod m$$

have a common solution x if and only if $a \equiv b \bmod \text{hcf}(n, m)$.

Exercise 1.13. Let R and S be rings. Show that the cartesian product $R \times S$ is a ring, with addition and multiplication defined by

$$(x, y) + (x', y') = (x + x', y + y'), \quad (x, y)(x', y') = (xx', yy').$$

Show that the map $\mathcal{C} : \mathbb{Z}/nm \to \mathbb{Z}/n \times \mathbb{Z}/m$ is a ring isomorphism.

1.6 PRIME NUMBERS

A positive integer p is called a *prime number* if $p \geq 2$ and the only factors of p are ± 1 and $\pm p$. The first few prime numbers are

$$2, \ 3, \ 5, \ 7, \ 11, \ 13, \ 17, \ldots$$

An integer $n > 1$ which is not prime is called a *composite* number. In the case that p is a prime number, we'll use the notation $\mathbb{F}_p$ for the ring $\mathbb{Z}/p$. This is because of the following useful fact.

Theorem 1.2. *If p is prime, then $\mathbb{F}_p$ is a field.*

Proof. We already know that $\mathbb{F}_p$ is a ring. To show that it is a field, we need to check two more axioms:

(1) $1 \not\equiv 0 \bmod p$,
(2) Every non-zero element of $\mathbb{F}_p$ is invertible.

The first of these is obviously true because $p \geq 2$. It remains to check that every non-zero element of $\mathbb{F}_p$ has an inverse. The non-zero elements of

$\mathbb{F}_p$ are $\{1, 2, \ldots, p-1\}$. None of these are multiples of p, so they must be coprime to p. By Proposition 1.2, they are invertible modulo p. □

Exercise 1.14. Write down a multiplication table for $\mathbb{F}_7^\times$ and find the inverse of each element.

Exercise 1.15. Let n be a positive integer. Show that if n is not prime, then $\mathbb{Z}/n$ is not a field (and so Theorem 1.2 can be made into an *if and only if* statement).

The prime numbers are thought of as the building blocks from which all integers are made. This is because every positive integer can be written as a product of primes. Indeed, if n is not itself prime, then it may be factorized into two smaller numbers; if they are not prime, then they may both be factorized, etc. This factorizing process cannot continue forever, since the factors get smaller with each step, so eventually we end up with a factorization of n into prime numbers. It turns out that this factorization into prime numbers is unique, even though we may have made many choices along the way to obtain the factorization.

Theorem 1.3 (Fundamental Theorem of Arithmetic). *For every positive integer n, there is a factorization*[a] $n = p_1 \cdots p_r$ *with prime numbers* $p_1, \ldots, p_r$. *If we have another factorization* $n = q_1 \cdots q_s$ *into prime numbers* q_i, *then* $r = s$ *and the primes* q_i *may be reordered so that* $p_i = q_i$ *for all* i.

The two sentences in the Fundamental Theorem of Arithmetic are called the *existence* and the *uniqueness* statements. The existence statement follows from the discussion above. However, the uniqueness statement is a little harder, and to prove this part, we shall use the following lemma.

Lemma 1.3 (Euclid's Lemma). *Let p be a prime number and let $a, b \in \mathbb{Z}$. If $p|ab$, then $p|a$ or $p|b$.*

Proof. The assumption that p divides ab means that $ab \equiv 0 \bmod p$. If $a \not\equiv 0 \bmod p$, then by Theorem 1.2, we know that a is invertible modulo p, and therefore, $b \equiv a^{-1} \times 0 \equiv 0 \bmod p$. Hence, either a or b is a multiple of p. □

Proof of the Fundamental Theorem of Arithmetic (*Existence*). We'll prove by induction on n. If $n = 1$, then the theorem is true with $r = 0$.

[a]It is implicit in this notation that when $n = 1$, we have $r = 0$, and the empty product on the right-hand side is equal to 1.

Assume that all positive integers less than n factorize into primes. Either n is prime, in which case, it is a product of exactly one prime, or $n = ab$ with a and b less than n. By the inductive hypothesis, a and b can be factorized into primes, and therefore so can n.

(*Uniqueness*) We'll prove again by induction on n. The uniqueness statement is clearly true for $n = 1$. Assume that all positive integers less than n have a unique factorizaion into primes, and suppose we have two factorizations of n

$$n = p_1 \cdots p_r = q_1 \cdots q_s$$

with p_i and q_i all prime. Clearly, p_1 is a factor of n, so by Euclid's Lemma, $p_1 | q_j$ for some j. After reordering the q_j's, we can assume that $p_1 | q_1$. Since q_1 is prime, we actually have $p_1 = q_1$. Hence,

$$\frac{n}{p_1} = p_2 \cdots p_r = q_2 \cdots q_s.$$

By the inductive hypothesis, $\frac{n}{p_1}$ has a unique factorization into primes. It follows that $r = s$ and we can reorder so that $p_i = q_i$ for all i. Therefore, the factorization of n is unique. $\square$

The Fundamental Theorem of Arithmetic never seems to surprise anyone. This is perhaps because we are so familiar with all the prime factorizations of small numbers that we are already aware of the uniqueness of prime factorizations, perhaps without being completely sure why they are unique. It's worth pointing out that the uniqueness of factorization is quite a special property of the ring $\mathbb{Z}$. One can easily come up with other rings in which factorization is not unique. For example, let

$$\mathbb{Z}[\sqrt{-5}] = \{x + y\sqrt{-5} : x, y \in \mathbb{Z}\}.$$

It's easy to see that $\mathbb{Z}[\sqrt{-5}]$ is closed under addition, subtraction and multiplication, so $\mathbb{Z}[\sqrt{-5}]$ is a subring of $\mathbb{C}$ (see Exercise 1.16). The number 6 in this ring factorizes in two very different ways.

$$6 = 2 \times 3 = (1 + \sqrt{-5})(1 - \sqrt{-5}).$$

From this, we see that $\mathbb{Z}[\sqrt{-5}]$ does not have unique factorization.

Exercise 1.16. Let R be a ring and let S be a subset of R satisfying the following conditions:

(1) S contains the elements 0 and 1 of the ring R;
(2) if x and y are in S, then $x + y$, $x - y$ and xy are also in S.

Show that S is a ring, with the operations of addition and multiplication from the ring R. (In these circumstances, S is called a *subring* of R.)

Trial division. It is clear, even from the existence part of the Fundamental Theorem of Arithmetic, that if n is composite, then the smallest prime factor p of n is no greater than $\sqrt{n}$. In practice, this makes it quite easy to find out whether relatively big numbers are prime. One simply lists the primes less than or equal to $\sqrt{n}$; if none of these are factors of n, then n is prime. This method is called *trial division.*

Example. We'll determine whether 199 is prime by trial division. The number 199 is between $14^2 = 196$ and $15^2 = 225$, so $\sqrt{199} < 15$. The primes up to 15 are $2, 3, 5, 7, 11, 13$, so we just need to check whether any one of these divides 199.

We can immediately see that 199 is not a multiple of 2, 3 or 5. For the other primes, we simply divide to check:

$$\begin{aligned} 199 &= 140 + 49 + 7 + 3 \equiv 3 && \mod 7, \\ 199 &= 99 + 99 + 1 \equiv 1 && \mod 11, \\ 199 &= 130 + 65 + 4 \equiv 4 && \mod 13. \end{aligned}$$

We've shown that 199 is not a multiple of any of the primes $\le \sqrt{199}$, so 199 is a prime number.

The time taken to show that an integer n is prime by trial division grows faster than $\sqrt{n}$ as n gets large (we need to divide by roughly $\frac{\sqrt{n}}{\log(n)}$ smaller primes, and each division takes time proportional to $\log(n)^2$). There are more sophisticated algorithms which test much more quickly whether a very large integer n is prime. For example, it's known that we can test for primality in time proportional to $(\log n)^6$ (see [1]).

Exercise 1.17. List the prime numbers up to 30. Hence, use trial division to determine which of the following numbers are prime and factorize those which are composite into primes.

$$263, \quad 323, \quad 329, \quad 540, \quad 617, \quad 851.$$

We can check using `sage` that we have the right answer using the command `is_prime`:

```
is_prime(199)
```

True

We can list all the primes between two numbers with the command `prime_range`:

```
prime_range(100,150)
```

[101, 103, 107, 109, 113, 127, 131, 137, 139, 149]

We can also find the factorization of a number into primes using the `factor` command.

```
factor(100)
```

$2^2 \cdot 5^2$

We'll next consider the question of how many prime numbers there are.

Theorem 1.4. *There are infinitely many prime numbers.*

Proof. We'll prove the theorem by contradiction. Suppose there are only finitely many primes, and let $p_1, \ldots, p_r$ be a complete list of all the prime numbers. Let $N = p_1 \cdots p_r + 1$. Clearly, N is congruent to 1 modulo every prime, so in particular, N is not a multiple of any prime number. This contradicts the Fundamental Theorem of Arithmetic. □

There are many variations on the proof of Theorem 1.4 given above, which show that there are infinitely many prime numbers of a particular kind. For example, one can show that there are infinitely many prime numbers congruent to 2 modulo 3. To see this, we shall argue by contradiction. Suppose for a moment that there were only finitely many such primes and call these primes $p_1, \ldots, p_r$. Consider the prime factors of the number

$$N = 3p_1 \cdots p_r - 1.$$

None of the primes $p_1, \ldots, p_r$ are factors of N, so N has no prime factors congruent to 2 modulo 3. Also, the number N is not a multiple of 3, so all of its prime factors are congruent to 1 modulo 3. By the Fundamental Theorem of Arithmetic, N is a product of primes which are congruent to 1 modulo 3, and therefore, $N \equiv 1 \bmod 3$. However, it's clear from the formula for N that $N \equiv 2 \bmod 3$, so we have a contradiction.

Here are some similar examples.

Exercise 1.18. Prove that there are infinitely many primes which are not congruent to 1 modulo 5.

Exercise 1.19. Let H be a proper subgroup[b] of the group $(\mathbb{Z}/n)^\times$. Prove that there are infinitely many prime numbers, whose congruency class modulo n is not in H.

Exercise 1.20. Show that there are no primes congruent to 210 modulo 1477.

Exercise 1.21. Show that there are infinitely many prime numbers p such that the congruence $x^5 + x + 1 \equiv 0 \bmod p$ has a solution.

Generalizing these exercises, one might ask which arithmetic progressions contain infinitely many prime numbers. Suppose we have a sequence

$$x_n = a + dn, \quad n \in \mathbb{N},$$

and we'd like to know whether the sequence contains infinitely many primes numbers. It's clear that $\text{hcf}(a, d)$ is a factor of every term x_n, and so the sequence can only contain more than one prime if a and d are coprime. It turns out that this is the only condition required for such a sequence to contain infinitely many prime numbers. In other words, we have the following.

Theorem 1.5 (Dirichlet's Primes in Arithmetic Progressions Theorem). *If a and d are coprime integers, then there are infinitely many prime numbers congruent to a modulo d.*

The proof of Dirichlet's Primes in Arithmetic Progressions Theorem goes beyond the scope of this book. Dirichlet's proof, which requires some knowledge of complex analysis, can be found in [6]. A more elementary (although not necessarily easier) proof was found much later by Selberg in 1949 [22].

One can also ask roughly how many prime numbers there are up to a given bound N, in other words, roughly how big is the quantity

$$\pi(N) = |\{p \leq N : p \text{ is a prime number}\}|.$$

It's useful when discussing $\pi(N)$ to use the notation of asymptotic equivalence. We say that two functions $f(N)$ and $g(N)$ are *asymptotically equivalent* (written $f(N) \sim g(N)$) if $\lim_{N\to\infty} \frac{f(N)}{g(N)} = 1$. By studying tables

[b]A *proper subgroup* of $(\mathbb{Z}/n)^\times$ is a subgroup which does not contain all the elements of $(\mathbb{Z}/n)^\times$.

of prime numbers, both Gauss and Legendre independently conjectured in the 1790s that $\pi(N)$ is approximately $\frac{N}{\log(N)}$.[c] Their conjecture was eventually made precise and proved roughly 100 years later independently by both Hadamard and de la Vallée–Poussin in 1896. The result is now called the *Prime Number Theorem.*

Theorem 1.6 (Prime Number Theorem). $\pi(N) \sim \frac{N}{\log(N)}$.

In fact, a slightly better estimate for $\pi(N)$ is the *logarithmic integral* $\text{Li}(N)$, defined by

$$\text{Li}(N) = \int_2^N \frac{1}{\log(x)} dx.$$

The logarithmic integral is asymptotically equivalent to $\frac{N}{\log(N)}$, but $|\pi(N) - \text{Li}(N)|$ is smaller than $|\pi(N) - \frac{N}{\log(N)}|$. It's an important open question to determine just how small $\pi(N) - \text{Li}(N)$ is. It is believed that the following statement is true (which is much stronger than the Prime Number Theorem).

Conjecture (Riemann Hypothesis). *There is a real number B such that for all $N \geq 2$,*

$$|\pi(N) - \text{Li}(N)| \leq B\sqrt{N}\log(N).$$

The Riemann Hypothesis is usually stated in a different way (see for example [17]), but the statement above is equivalent.

Exercise 1.22. Let $a > 2$ and $b \geq 2$ be integers, such that $a^b - 1$ is a prime number. Show that $a = 2$ and b is prime.

The prime numbers of the form $2^p - 1$ are called *Mersenne Primes*. Not all numbers of the form $2^p - 1$ are prime, but it has been conjectured that there are infinitely many Mersenne Primes.

Exercise 1.23. Let $k \geq 2$ be an integer such that $2^k + 1$ is a prime number. Show that $k = 2^n$ for some integer n.

Numbers of the form $F_n = 2^{2^n} + 1$ are called *Fermat numbers*. This is because Fermat wrongly conjectured that all such numbers are prime. In fact, the only known Fermat prime numbers are F_0, F_1, F_2, F_3 and F_4. It is not known whether there are infinitely many prime numbers of the

[c] We write log for the logarithm to the base e.

form F_n, or even whether there are infinitely many composite numbers of this form.

Exercise 1.24. Find all solutions to the simultaneous congruences $x \equiv 5 \bmod 2^{400}$ and $x \equiv 5 \bmod 3^{400}$.

HINTS FOR SOME EXERCISES

1.5. Use Bezout's lemma to show that $\mathrm{hcf}(n,m)\mathrm{lcm}(n,m)$ is a multiple of nm and prove directly from the definition that $\mathrm{lcm}(n,m) \leq \frac{nm}{\mathrm{hcf}(n,m)}$.

1.12. This is an if and only if statement, so there are two things to prove. If we assume that the congruences have solutions, then we can prove directly that $a \equiv b \bmod \mathrm{hcf}(n,m)$. To prove the converse, it's convenient to rewrite the congruences in terms of the variable $y = \frac{x-a}{\mathrm{hcf}(n,m)}$ and use the Chinese Remainder Theorem to show that an integer solution y exists.

1.18. $N = 5p_1 \cdots p_r - 1$.

1.22. Show that if b is a factor of c, then $a^b - 1$ is a factor of $a^c - 1$.

1.23. Show that if $k = rs$ with r odd, then $2^s + 1$ is a factor of $2^k + 1$.

Chapter 2

Polynomial Rings

In this chapter, we shall study polynomials. At first sight, polynomials seem quite different from integers. However, the methods we shall use to study polynomials are very similar to the methods we used in Chapter 1 to study integers. In particular, we'll see that for any field $\mathbb{F}$, the polynomial ring $\mathbb{F}[X]$ has a Euclidean algorithm. As a consequence, we are able to prove versions for $\mathbb{F}[X]$ of many of the results of Chapter 1. In this sense, the material covered in this chapter fits rather well together with the material covered in Chapter 1.

The second reason for studying polynomials at this point is because some of the more advanced theorems concerning $\mathbb{Z}$, which we'll see in Chapter 3, are most naturally proved using properties of polynomial rings.

Notation. Let R be a ring. A *polynomial* over R is an expression of the form

$$f(X) = a_0 + a_1 X + \cdots + a_n X^n,$$

with coefficients $a_0, \dots, a_n$ in R. The set of all polynomials over R is written as $R[X]$. We can add and multiply polynomials in the usual way. The operations of addition and multiplication on $R[X]$ satisfy the Ring Axioms, and so $R[X]$ is itself a ring.

Remark. There is one subtle point in the definition of a polynomial. When we talk about a polynomial, we're talking about the expression itself, i.e. the sequence of coefficients $a_0, \dots, a_n$. We do not simply regard polynomials as functions of a variable X. The reason for this is because it is possible for two distinct polynomials to take the same values for all $X \in R$. For example, suppose R is the field $\mathbb{F}_2 = \{0, 1\}$. Then the polynomials X^2 and X take the same values for all $X \in \mathbb{F}_2$, although they are not regarded as the same polynomial. There will always be such examples when R is a finite ring: this is because there are only finitely many functions from R to R, but there are infinitely many distinct polynomials. However, if $\mathbb{F}$ is an infinite field, then two distinct polynomials will always give rise to two distinct functions on $\mathbb{F}$ (see Exercise 2.2).

2.1 LONG DIVISION OF POLYNOMIALS

The concepts of *factor* and *common factor* introduced in Chapter 1 make sense in any ring. In this section, we shall consider these concepts in the ring $\mathbb{F}[X]$ of polynomials over a field $\mathbb{F}$. For two polynomials f and g in $\mathbb{F}[X]$, we say that g is a factor of f (written $g|f$) if there is a polynomial $h \in \mathbb{F}[X]$ such that $f = g \times h$. We'll see that there is a version of the Euclidean algorithm which works in the ring $\mathbb{F}[X]$ and allows us to calculate the "highest common factor" of two polynomials.

The degree of a polynomial. Consider a polynomial f over a field $\mathbb{F}$, defined as follows:

$$f(X) = a_0 + a_1X + \cdots + a_nX^n, \quad a_0, \ldots, a_n \in \mathbb{F}.$$

The largest integer d, for which the coefficient a_d is non-zero, is called the *degree* of f and is written as $\deg(f)$. For example, the degree of $0X^4+2X^3+0X^2-4X+1$ is 3. It is convenient to define the degree of the zero polynomial to be $-\infty$. Polynomials of degree ≤ 0 are called *constant polynomials.* If d is the degree of f, then the term a_dX^d is called the *leading term* of $f(X)$ and a_d is called the *leading coefficient*. The coefficient a_0 is called the *constant term* of f. A polynomial is called *monic* if its leading coefficient is 1.

Proposition 2.1. *If f and g are non-zero polynomials over a field, then their product is non-zero and*

$$\deg(fg) = \deg(f) + \deg(g). \tag{2.1}$$

Proof. If f has a leading term aX^n and g has a leading term bX^m, then the product fg will have a leading term abX^{n+m}. Note that the coefficent ab is not zero, since a and b are non-zero elements of the field $\mathbb{F}$. □

In fact, Equation (2.1) is true even when f or g is zero, as long as we use the convention that $-\infty + n = -\infty$ for every natural number n.

Corollary 2.1. *Suppose f, g and h are polynomials over a field, with $f \neq 0$. If $fg = fh$, then $g = h$.*

Proof. Since $f \times (g - h) = 0$ and $f \neq 0$, it follows from Proposition 2.1 that $g - h = 0$, so $g = h$. $\square$

Corollary 2.1 tells us that if f is a factor of g, then there is a *unique* polynomial h such that $g = fh$. In view of this, it makes sense to write $\frac{g}{f}$ for this polynomial.

Corollary 2.2. *The invertible elements in the ring* $\mathbb{F}[X]$ *are exactly the non-zero constant polynomials.*

Proof. Let f be a unit with inverse g, so that $fg = 1$. Then $\deg f + \deg g = \deg(1) = 0$ by Proposition 2.1. Therefore, $\deg f = 0$. $\square$

Note that the previous statements rely on $\mathbb{F}$ being a field and are false in some other cases. For example in the ring $(\mathbb{Z}/4)[X]$, we have $(1+2X)(1+2X) = 1$, and so $1 + 2X$ is invertible, with inverse $1 + 2X$.

Highest common factors. Let f and g be polynomials in $\mathbb{F}[X]$ which are not both zero. A polynomial $h \in \mathbb{F}[X]$ is called a common factor of f and g if $h|f$ and $h|g$. A common factor h is called a *highest common factor* if h is monic and every common factor of f and g has degree $\leq \deg h$.

It's fairly obvious that f and g will have at least one highest common factor. Indeed, if we take any common factor d of the largest possible degree, then $c^{-1}d$ is a highest common factor, where c is the leading coefficient of d. We'll see in Proposition 2.4 that f and g cannot have two different highest common factors.

Long Division of Polynomials. To calculate the highest common factor of two polynomials f and g, we use a method similar to that used for calculating highest common factor integers. The idea is that we divide f by g with remainder, so that we have an equation of the form

$$f = qg + r, \quad q, r \in \mathbb{F}[X].$$

The common factors of f and g are the same as the common factors of g and r. Therefore, the highest common factors of f and g are the same as the highest common factors of g and r. This allows us to calculate the highest common factors by a sequence of divisions with remainder, just as we did in $\mathbb{Z}$ in Chapter 1.

To divide polynomials with remainder, we use a method called the *long division* algorithm because of its similarity to long division of integers. To explain how this algorithm works, let's suppose we have polynomials f and g over a field $\mathbb{F}$ with $g \neq 0$. We'll define a sequence of polynomials $f_1, \ldots, f_n$

beginning with $f_1 = f$. Suppose the leading term of f_1 is aX^d, and let's assume that the leading term of g is bX^e. If $\deg f_1 \geq \deg g$, then $d - e \geq 0$, and the leading term of $\frac{a}{b}X^{d-e}g(X)$ is the same as that of f_1. This means that if we subtract $\frac{a}{b}X^{d-e}g(X)$ from f_1, then we will obtain a polynomial f_2 of smaller degree than f_1. As long as the degree of f_2 is at least as big as that of g, we may subtract a multiple of g from f_2 to obtain a polynomial f_3 of smaller degree than f_2. Continuing in this way, we eventually reach a polynomial f_n whose degree is smaller than that of g, at which point we can go no further. In this way, we obtain a sequence of polynomials $f = f_1, f_2, \ldots, f_n$, each of smaller degree than the previous one, such that the final polynomial f_n has smaller degree than g. Each f_{r+1} differs from f_r by a multiple of g, and therefore, $f_n - f_1$ is a multiple of g. If we define $q = \frac{f_1 - f_n}{g}$ and $r = f_n$, then we have (analogously to Equation (1.1))

$$f = qg + r, \quad \deg r < \deg g.$$

To sum things up, we have proved the following.

Proposition 2.2. *Let f and g be polynomials over a field $\mathbb{F}$ with $g \neq 0$. Then there exist polynomials q and r over $\mathbb{F}$ such that*

$$f = qg + r, \quad \text{and} \quad \deg(r) < \deg(g).$$

Example. Consider the following polynomials over $\mathbb{Q}$:

$$f(X) = X^3 + X^2 + 2X + 1, \quad g(X) = 2X + 1.$$

We shall divide f by g, following the algorithm described above.

(1) Let $f_1 = f$ and divide the leading term X^3 of $f_1(X)$ by the leading term $2X$ of $g(X)$. This gives us $\frac{1}{2}X^2$.

$$\begin{array}{r|llll} & \frac{1}{2}X^2 & & & \\ \cline{2-5} 2X+1 & X^3 & +X^2 & +2X & +1 \end{array}.$$

(2) The product $\frac{1}{2}X^2 g(X)$ has the same leading coefficient as f_1. We subtract this product from $f_1(X)$ to obtain a polynomial $f_2(X) = \frac{1}{2}X^2 + 2X + 1$ of smaller degree than f_1.

$$\begin{array}{r|llll} & & \frac{1}{2}X^2 & & \\ \hline 2X+1 & X^3 & +X^2 & +2X & +1 \\ & X^3 & +\frac{1}{2}X^2 & & \\ \cline{2-5} & & \frac{1}{2}X^2 & +2X & +1 \end{array}.$$

So far we've shown that $f_1(X) = (\frac{1}{2}X^2)g(X) + f_2(X)$. We can continue with the algorithm because $\deg f_2 \geq \deg g$.

(3) We repeat the process, this time dividing the leading term of $f_2(X) = \frac{1}{2}X^2 + 2X + 1$ by the leading term of $g(X)$.

$$\begin{array}{r|llll} & & \frac{1}{2}X^2 & +\frac{1}{4}X & \\ \hline 2X+1 & X^3 & +X^2 & +2X & +1 \\ & X^3 & +\frac{1}{2}X^2 & & \\ \cline{2-5} & & \frac{1}{2}X^2 & +2X & +1 \\ & & \frac{1}{2}X^2 & +\frac{1}{4}X & \\ \cline{3-5} & & & \frac{7}{4}X & +1 \end{array}.$$

This shows that $f_1(X) = (\frac{1}{2}X^2 + \frac{1}{4}X)g(X) + (\frac{7}{4}X + 1)$. Again, the degree of the remainder $f_3(X) = \frac{7}{4}X + 1$ is no smaller than that of g.

(4) Finally, we divide the leading term of $f_3(X) = \frac{7}{4}X + 1$ by the leading term of $g(X)$.

$$\begin{array}{r|llll} & & \frac{1}{2}X^2 & +\frac{1}{4}X & +\frac{7}{8} \\ \hline 2X+1 & X^3 & +X^2 & +2X & +1 \\ & X^3 & +\frac{1}{2}X^2 & & \\ \cline{2-5} & & \frac{1}{2}X^2 & +2X & +1 \\ & & \frac{1}{2}X^2 & +\frac{1}{4}X & \\ \cline{3-5} & & & \frac{7}{4}X & +1 \\ & & & \frac{7}{4}X & +\frac{7}{8} \\ \cline{4-5} & & & & \frac{1}{8} \end{array}.$$

Our remainder is now $f_4(X) = \frac{1}{8}$, which has smaller degree than g, so we stop at this point. We've shown that $f = qg + r$, with the quotient q and remainder r given by

$$q(X) = \frac{1}{2}X^2 + \frac{1}{4}X + \frac{7}{8}, \quad r(X) = \frac{1}{8}.$$

Calculations of the kind described above can be carried out in `sage`. In order to do this, we first need to define an appropriate ring of polynomials.

`var('X')`	let `X` be a new variable.
`R.<X>=QQ[X]`	`R` is defined to be the polynomial ring $\mathbb{Q}[X]$.

Having defined the ring of polynomials, we can now do the calculation from the example above, where we divide $X^3 + X^2 + 2X + 1$ by $2X + 1$.

`f,g= X^3+X^2+2*X+1, 2*X+1`	define the polynomials f and g in $\mathbb{Q}[X]$.
`f % g`	the remainder on dividing f by g
$\frac{1}{8}$	
`f.quo_rem(g)`	This returns the quotient and remainder obtained on dividing f by g.
$\left(\frac{1}{2}X^2 + \frac{1}{4}X + \frac{7}{8}, \frac{1}{8}\right)$	

The long division algorithm relies on the fact that $\mathbb{F}$ is a field and will not usually work with polynomials over other rings. For example, in the ring $\mathbb{Z}[X]$, we may still divide with remainder by the same method. However, as we saw in the example above, neither the quotient nor the remainder need be in $\mathbb{Z}[X]$. If we look again at the long division algorithm, we see why this is the case. In order to reduce the degree of the polynomial f_1, we subtracted $b^{-1}aX^{d-e}g(X)$, where b is the leading coefficient of g. If $\mathbb{F}$ were not a field, then there is no reason why there should be an element b^{-1}. If however g is a *monic* polynomial, then $b = 1$, and we instead subtract $aX^{d-e}g(X)$ from f_1. This gives us the following generalization of Proposition 2.2.

Proposition 2.3. *Let R be an arbitrary ring and let $f, g \in R[X]$ with g monic. Then there exist polynomials $q, r \in R[X]$ such that $f = qg + r$ and* $\deg(r) < \deg(g)$.

Roots of Polynomials. The long division algorithm gives us a way of checking whether a polynomial g is a factor of another polynomial f: we divide f by g and just check whether the remainder is zero. In the case that

g has degree 1, this method can be simplified by the following result (which we prove over an arbitrary ring).

Theorem 2.1 (The Remainder Theorem). *Let α be an element of a ring R and let f be a polynomial over R. Then $X - \alpha$ is a factor of f if and only if $f(\alpha) = 0$.*

Proof. Suppose first that $X - \alpha$ is a factor of f, so we have $f(X) = (X - \alpha)g(X)$ for some polynomial g. This implies $f(\alpha) = (\alpha - \alpha)g(\alpha) = 0$.

Conversely, suppose $f(\alpha) = 0$. By Proposition 2.3, we can divide f by $X - \alpha$ in the ring $R[X]$ with remainder:

$$f(X) = q(X)(X - \alpha) + r(X), \quad \deg(r) < 1. \tag{2.2}$$

To show that $X - \alpha$ is a factor of f, we need to show that the remainder r is zero. Since r has a degree less than 1, it is a constant in R rather than a polynomial. We can calculate the value of r by substituting $X = \alpha$ into Equation (2.2). This gives us $r = f(\alpha) - q(\alpha)(\alpha - \alpha) = 0$. Hence, $f(X) = q(X)(X - \alpha)$, so $X - \alpha$ is a factor of f. □

Corollary 2.3. *Let f be a non-zero polynomial of degree d over a field $\mathbb{F}$. Then there are at most d elements $\alpha \in \mathbb{F}$, such that $f(\alpha) = 0$.*

Proof. The proof is by induction on d. If $d = 0$, then f is a non-zero constant, and therefore, f has no roots. Next, assume that $d > 0$, and that polynomials of degree $d-1$ have no more than $d-1$ roots. If f has no roots, then the corollary is true for f, so we may assume that f has at least one root $\alpha \in \mathbb{F}$. By the Remainder Theorem, $f(X) = (X - \alpha)g(X)$ for some polynomial g of degree $d - 1$. By the inductive hypothesis, g has no more than $d - 1$ roots in $\mathbb{F}$. Furthermore, since $\mathbb{F}$ is a field, every root of f is either equal to α or a root of g. Hence, f has no more than d roots. □

Exercise 2.1. Let $f(X) = X^5 + X$ and $g(X) = X^2 + X + 1$ over the field $\mathbb{F}_2$. Find $q, r \in \mathbb{F}_2[X]$ with $\deg r < 2$, such that $f = qg + r$.

Exercise 2.2. Let $\mathbb{F}$ be an infinite field, and suppose f and g are polynomials in $\mathbb{F}[X]$ such that $f(a) = g(a)$ for every $a \in \mathbb{F}$. Show that $f = g$.

Exercise 2.3. Let $f(X)$ be a polynomial with coefficients in $\mathbb{Z}$, and for any $n > 0$, let $a(n)$ be the number of solutions $x \in \mathbb{Z}/n$ to the congruence

$$f(x) \equiv 0 \bmod n.$$

(1) Show that if $\text{hcf}(n, m) = 1$, then $a(nm) = a(n)a(m)$.
(2) How many solutions does the following congruence have?

$$x^2 \equiv 1 \bmod (2 \cdot 3 \cdot 5 \cdot 7 \cdot 11 \cdot 13 \cdot 17 \cdot 19).$$

Exercise 2.4. Let $\mathbb{F}$ be a field contained in a larger field $\mathbb{E}$, and let f and g be polynomials over $\mathbb{F}$. Show that f is a factor of g in the ring $\mathbb{F}[X]$ if and only if f is a factor of g in $\mathbb{E}[X]$.

2.2 HIGHEST COMMON FACTORS

In this section, we shall consider polynomials over a field $\mathbb{F}$. Recall that the highest common factor of two polynomials over $\mathbb{F}$ is a monic polynomial of largest possible degree, which is a factor of both polynomials.

When we were considering highest common factors in the ring $\mathbb{Z}$ in Chapter 1, we proved results by induction on the smaller of the numbers n and m. To make the inductive step work, we divided n by m with remainder to obtain an expression of the form

$$n = qm + r, \quad 0 \leq r < m.$$

We then observed that the common factors of n and m are the same as the common factors of m and r, and so $\text{hcf}(n, m) = \text{hcf}(m, r)$. We shall use the same idea in the ring $\mathbb{F}[X]$. If f and g are polynomials with $g \neq 0$, then using the long division algorithm, we obtain an expression of the form

$$f = qg + r, \quad \deg r < \deg g.$$

Again, it is clear that the common factors of f and g are exactly the same as the common factors of g and r. This allows us to prove results concerning highest common factors by induction on the degree of a polynomial. The first example of such a proof is the following important fact.

Proposition 2.4. *If f and g are polynomials over a field $\mathbb{F}$ which are not both zero, then f and g have a unique highest common factor in $\mathbb{F}[X]$.*

Remark. The question of uniqueness of highest common factors does not arise in $\mathbb{Z}$; clearly, if we have two natural numbers, then one of them is the biggest. This question does however arise with the ring $\mathbb{F}[X]$, since there are many monic polynomials with the same degree.

Proof. We'll prove the result by induction on the degree of g. The start of the induction will be the case that g is the zero polynomial, which has degree $-\infty$. Suppose we have two highest common factors, which we'll call

h_1 and h_2; we must show that $h_1 = h_2$. In this case, f is itself a common factor of f and g, and there is no factor of f of larger degree than f. Hence, h_1 and h_2, being highest common factors, must have the same degree as f. As they are both factors of f, we have $f = c_1h_1 = c_2h_2$ with non-zero constants $c_1, c_2 \in \mathbb{F}^\times$. Since highest common factors are assumed to be monic, it follows that c_1 and c_2 are both equal to the leading coefficient of f, and $h_1 = h_2$.

For the inductive step, we suppose that g is non-zero and we assume the proposition is true for any pair of polynomials, one of which has smaller degree than g. We may divide f by g with remainder to obtain an expression of the form

$$f = qg + r, \quad \deg r < \deg g.$$

Since r has smaller degree than g, the inductive hypothesis tells us that g and r have a unique highest common factor. However, the common factors of f and g are exactly the same as the common factors of g and r. Therefore, f and g also have a unique highest common factor. □

From now on, we'll write $\text{hcf}(f, g)$ for the highest common factor of two polynomials f and g. Proposition 2.4 guarantees that there is only one highest common factor, and so the notation is not ambiguous. As in the ring $\mathbb{Z}$, the long division algorithm gives us a method for calculating highest common factors. Suppose we have polynomials f and g with $g \neq 0$. If we divide f by g, then we obtain an expression of the form

$$f = qg + r,$$

where q and r are polynomials and r has a smaller degree than g. As we noted above, this implies

$$\text{hcf}(f, g) = \text{hcf}(g, r). \tag{2.3}$$

To calculate the highest common factor of f and g, we construct a sequence of polynomials $r_1, \ldots, r_n, r_{n+1} = 0$ as follows:

(1) We set $r_1 = f$ and $r_2 = g$.
(2) If $r_s \neq 0$, then r_{s+1} is defined to be the remainder on dividing r_{s-1} by r_s. Note that r_{s+1} has a smaller degree than r_s. Therefore, the sequence of polynomials must eventually end with a remainder of 0.

After applying Equation (2.3) several times, we find that

$$\text{hcf}(f, g) = \text{hcf}(r_1, r_2) = \text{hcf}(r_2, r_3) = \cdots = \text{hcf}(r_n, 0) = c^{-1}r_n,$$

where c is the leading coefficient of r_n.

Example. Consider the following polynomials over the field $\mathbb{F}_2$:

$$f(X) = X^3 + X + 1, \quad g(X) = X^2 + X + 1.$$

We'll calculate $\operatorname{hcf}(f, g)$ by the method described above. As in Chapter 1, we'll write the sequence r_n in bold. Arithmetic is rather easy in $\mathbb{F}_2[X]$ because addition is the same as subtraction and all non-zero polynomials are monic.

$$\begin{aligned} \mathbf{f(X)} &= (X+1)\mathbf{g(X)} + \mathbf{X}, \\ \mathbf{g(X)} &= (X+1)\mathbf{X} + \mathbf{1}, \\ \mathbf{X} &= X \times \mathbf{1} + 0. \end{aligned}$$

By Equation (2.3), we have

$$\operatorname{hcf}(f, g) = \operatorname{hcf}(g, X) = \operatorname{hcf}(X, 1) = \operatorname{hcf}(1, 0) = 1.$$

Lemma 2.1 (Bezout's Lemma for Polynomials). *Let f and g be polynomials over a field $\mathbb{F}$, which are not both zero. Then there exist polynomials h and k in $\mathbb{F}[X]$ such that*

$$\operatorname{hcf}(f, g) = hf + kg. \tag{2.4}$$

Proof. The proof is another example of a proof by induction on the degree. In the case $g = 0$, we have $\operatorname{hcf}(f, g) = c^{-1}f$, where c is the leading coefficient of f. Thus, the lemma is true with $h = c^{-1}$ and $k = 0$.

Let's now assume that g is not zero, and that the lemma is true for all pairs of polynomials, one of which has smaller degree than g. By Proposition 2.2, we may divide f by g with remainder:

$$f = qg + r, \quad \deg r < \deg g.$$

Since r has a smaller degree than g, the inductive hypothesis tells us that there exist polynomials u and v, such that

$$\operatorname{hcf}(g, r) = ug + vr.$$

On the other hand, since $\operatorname{hcf}(f, g) = \operatorname{hcf}(g, r)$, we can rewrite the equation as

$$\operatorname{hcf}(f, g) = vf + (u - vq)g.$$

This shows that $h = v$, $k = u - vq$ is a solution to Equation (2.4). This proves the lemma. □

Just as in the ring $\mathbb{Z}$, we may use Euclid's algorithm in $\mathbb{F}[X]$ to find polynomials h and k which solve Equation (2.4). This method is described for $\mathbb{Z}$ in Section 1.2.

Example. We return to the example considered above. Let f and g be the polynomials over $\mathbb{F}_2$ defined as follows:

$$f(X) = X^3 + X + 1, \quad g(X) = X^2 + X + 1.$$

We'll find a solution to Equation (2.4) in this case. We've shown that the highest common factor of f and g is 1. Next, we write the highest common factor in terms of f and g.

$$\begin{aligned} \mathbf{1} &= \mathbf{g(X)} + (X+1)\mathbf{X} \\ &= \mathbf{g(X)} + (X+1)\big(\mathbf{f(X)} + (X+1)\mathbf{g(X)}\big) \\ &= \big(1 + (X+1)^2\big)\mathbf{g(X)} + (X+1)\mathbf{f(X)} \\ &= X^2\mathbf{g(X)} + (X+1)\mathbf{f(X)}. \end{aligned}$$

We've shown that $h(X) = X + 1$ and $k(X) = X^2$ is a solution to Equation (2.4).

Exercise 2.5. Let $f(X) = 2X^4 + 5X^3 + 8X^2 + 7X + 4$ and $g(X) = 2X^2 + 3X + 3$ in $\mathbb{Q}[X]$.

(1) Show that $\mathrm{hcf}(f, g) = 1$.
(2) Find polynomials $h, k \in \mathbb{Q}[X]$ such that $hf + kg = 1$.
(3) Suppose $\alpha \in \mathbb{C}$ satisfies $f(\alpha) = 0$. Find $a, b, c, d \in \mathbb{Q}$ such that

$$\frac{1}{g(\alpha)} = a + b\alpha + c\alpha^2 + d\alpha^3.$$

Exercise 2.6. Let $f(X) = X^4 - X^3 - 3X^2 + X + 1$ and $g(X) = X^3 - 3X^2 + X + 1$ in $\mathbb{F}_5[X]$. Calculate the highest common factor of f and g and find polynomials $h, k \in \mathbb{F}_5[X]$ such that $hf + kg = \mathrm{hcf}(f, g)$.

One can check answers to these questions on `sage` using the `gcd` and `xgcd` commands in the same way as we did with integers in Chapter 1. For example, here is how to answer Exercise 2.6.

```
F=GF(5)                                  F is the field F_5
var('X')
R.<X>=F[X]                               R is the ring F_5[X]
f,g=X^4-X^3-3*X^2+X+1, X^3-3*X^2+X+1
gcd(f,g)                                 this returns hcf(f,g)
    1
xgcd(f,g)                                this returns (hcf(f,g),h,k),
    (1, 2X^2 + 4X + 3, 3X^3 + 2X^2 + 3)  where hf + kg = hcf(f,g).
```

Exercise 2.7. Let f and g be elements of $\mathbb{F}[X]$ which are not both zero. Show that every common factor of f and g is a factor of $\mathrm{hcf}(f,g)$.

Exercise 2.8. Let $\mathbb{F}$ be a field contained in a larger field $\mathbb{E}$, and let f and g be polynomials over $\mathbb{F}$. Prove that the highest common factor of f and g in the ring $\mathbb{F}[X]$ is equal to the highest common factor of f and g in the ring $\mathbb{E}[X]$.

2.3 UNIQUENESS OF FACTORIZATION

In this section, we introduce the idea of an *irreducible polynomial*, which is very much like the idea of a prime number. We'll see that every polynomial over a field may be factorized into irreducible polynomials and that such a factorization is essentially unique.

For a moment, let's suppose R is any ring. We may divide the elements of R into three classes:

1. The *invertible* elements, or *units* in R. These are the elements $x \in R$ for which there is an element $x^{-1} \in R$ satisfying $x \times x^{-1} = 1$.
2. The *reducible* elements. These are elements $x \in R$ which have a factorization $x = yz$, where neither y nor z is a unit.
3. The *irreducible* elements. These are the elements of R which are neither invertible nor reducible.

For example, in the ring $\mathbb{Z}$, the invertible elements are 1 and -1, and the irreducible elements are the numbers p and $-p$, where p is a prime number. We now consider the ring $\mathbb{F}[X]$ of polynomials over a field $\mathbb{F}$. We've seen in Corollary 2.2 that the invertible elements in $\mathbb{F}[X]$ are the non-zero constant polynomials. A non-constant polynomial is irreducible if cannot be factorized as a product of polynomials of smaller degrees.

Lemma 2.2 (Euclid's Lemma for Polynomials). *Let p be an irreducible polynomial over a field $\mathbb{F}$. Suppose $p|ab$ with polynomials $a, b \in \mathbb{F}[X]$. Then $p|a$ or $p|b$ (or both).*

Proof. We'll assume that p is not a factor of a and show that p must be a factor of b. As p is irreducible, the only monic factors of p are 1 and $c^{-1}p$, where c is the leading coefficient of p. By assumption, $c^{-1}p$ is not a factor of a, and so it follows that $\text{hcf}(a, p) = 1$. Hence, there exist polynomials h and k such that $1 = ha + kp$. Multiplying this equation by b, we get

$$b = hab + kpb.$$

We are assuming that p is a factor of ab and clearly p is a factor of kpb. Therefore, p is a factor of b. □

Theorem 2.2 (Uniqueness of factorization of polynomials). *Let f be a non-zero polynomial over a field $\mathbb{F}$. Then there is a unique factorization*

$$f = cp_1p_2 \cdots p_r,$$

where $c \in \mathbb{F}^\times$ is the leading coefficient of f and each factor p_i is monic and irreducible. (Here, "unique" means up to reordering the factors p_i.)

We include the proof for completeness, although it is essentially the same as the proof of the Fundamental Theorem of Arithmetic, which is the analogous statement for the ring $\mathbb{Z}$.

Proof. Without loss of generality, we shall assume that f is monic, so we must show that f can be written uniquely as a product of irreducible monic polynomials.

(*Existence*) The proof is by induction on the degree of f. If the degree of f is 0, then f is equal to 1, which is the empty product. Assume that $\deg(f) > 0$, and that all monic polynomials of a smaller degree than f may be factorized into irreducible, monic polynomials. If f is itself irreducible, then there is nothing to prove. If not, then there is a factorization $f = gh$, where g and h are monic polynomials of smaller degree than f. By the inductive hypothesis, both g and h are products of irreducible monic polynomials. Therefore, f is a product of irreducible monic polynomials.

(*Uniqueness*) The proof is by induction on the degree of f. Clearly, the constant polynomial 1 does not have two different factorizations. Let's assume that $\deg(f) > 0$ and that all monic polynomials of smaller degree than f have a unique factorization into irreducible monic polynomials. Suppose we have two factorizations of f:

$$f = p_1 \cdots p_r = q_1 \cdots q_s, \quad p_i, q_j \text{ irreducible and monic.}$$

Then p_1 is a factor of $q_1 \cdots q_s$, and so by Euclid's Lemma for Polynomials, it follows that p_1 must be a factor of one of the polynomials q_j. After reordering the polynomials q_j, we can assume that p_1 is a factor of q_1. As q_1 is irreducible and monic, we see that actually $p_1 = q_1$. This implies (using Corollary 2.1) that

$$p_2 \cdots p_r = q_2 \cdots q_s.$$

However, the polynomial $p_2 \cdots p_r$ has a smaller degree than f and so has a unique factorization into irreducible monic polynomials by the inductive hypothesis. This means that $r = s$, and we may reorder the polynomials so that $p_i = q_i$ for all i. □

Exercise 2.9. Let $\mathbb{F}$ be a field. Prove that there are infinitely many monic irreducible polynomials in $\mathbb{F}[X]$.

Exercise 2.10. Show that there are infinitely many irreducible polynomials p over $\mathbb{F}_2$ such that $p(0) = 1$ and $p'(0) = 1$.

2.4 IRREDUCIBLE POLYNOMIALS

We've shown that every polynomial over a field may be factorized uniquely into irreducible polynomials over that field. To make this statement a little more concrete, we next describe some ways of recognizing whether a polynomial is irreducible or not. Such questions are quite easy for some fields and more difficult for others. In number theory, we are mainly concerned with irreducible polynomials over $\mathbb{F}_p$ and $\mathbb{Q}$. However, we'll also describe how to recognize irreducible polynomials over the real and complex numbers, which is much easier.

Irreducible real and complex polynomials. We begin with a simple observation which is true over any field:

Every polynomial of degree 1 *is irreducible.*

In fact, when the field is $\mathbb{C}$, this statement gives us all the irreducible polynomials.

Theorem 2.3 (Fundamental Theorem of Algebra). *Let $f \in \mathbb{C}[X]$. Then f is irreducible if and only if f has degree* 1.

The proof of this theorem goes a little outside the scope of this book. One can currently find several proofs on the Wikipedia page. Alternatively, it is proved in most textbooks on complex analysis (see, for example, Theorem 10.7 in Chapter 10 of the book, *Complex Analysis* [20]). The first "proof" is due to Gauss in 1799 [9]; Gauss' proof used an intuitive geometric argument which is quite difficult to make rigorous. A more complete proof was found by Jean–Robert Argand in 1814 [2]. Ironically, Argand is far better known for the "Argand diagram", which was an idea of Gauss.

We next consider polynomials over the real numbers. It's clear that there are irreducible polynomials over $\mathbb{R}$ of degree bigger than 1. For example, the polynomial $X^2 + 1$ has no real roots. Hence, by the Remainder Theorem, $X^2 + 1$ has no factors of degree 1, so it must be irreducible. The next result tells us that this is essentially the only way in which we can obtain irreducible real polynomials of a degree bigger than 1.

Proposition 2.5. *The irreducible monic polynomials over $\mathbb{R}$ are as follows:*

(1) *the linear polynomials $X + a$ with $a \in \mathbb{R}$,*
(2) *the quadratic polynomials X^2+bX+c with $b^2-4c < 0$ (i.e. the quadratic polynomials with no real roots).*

The proof is in Exercise 2.11.

Irreducible polynomials over $\mathbb{F}_p$. Let p be a prime number and let f be a polynomial over the field $\mathbb{F}_p$ of integers modulo p. Checking whether f is irreducible is rather like checking whether an integer is prime. If f factorizes, then its smallest non-trivial factor has degree no more than $\frac{1}{2}\deg(f)$. There are only finitely many polynomials of any given degree, so there are only finitely many possible factors that we need to check.

Example. Consider the polynomials $f(X) = X^3 + 2X + 1$ and $g(X) = X^3 + 3X + 1$ over the field $\mathbb{F}_5$. Since f and g have degree 3, each of them factorizes if and only if it has a linear factor, and hence a root in $\mathbb{F}_5$. The values of f and g on $\mathbb{F}_5$ are given in the following table.

X	$f(X) = X^3 + 2X + 1 \bmod 5$	$g(X) = X^3 + 3X + 1 \bmod 5$
0	1	1
1	4	0
2	3	0
3	4	2
4	3	2

As none of the values of $f(X)$ are zero, f is irreducible. On the other hand, g has roots at $X = 1$ and $X = 2$, so g factorizes. In fact, by the Remainder Theorem, both $X - 1$ and $X - 2$ are factors of g, so we have

$$g(X) \equiv (X-1)(X-2)(X-a) \mod 5, \quad a \in \mathbb{F}_5.$$

Substituting $X = 0$ into this equation, we see that $1 \equiv -2a \bmod 5$. Solving this congruence, we get $a \equiv 2 \bmod 5$, and so the factorization of g into irreducibles is $g(X) = (X-1)(X-2)^2$.

Irreducible polynomials over $\mathbb{Q}$. It is rather harder to check whether a polynomial f over $\mathbb{Q}$ is irreducible than it is to check whether one is irreducible over $\mathbb{F}_p$. The reason for this is because there are infinitely many irreducible polynomials of any fixed degree, so one cannot simply check all the possible factors by the trial division method. However, the following result allows us to narrow down the list of possible factors. It shows that if we have a polynomial with integer coefficients, then it is enough to consider factorizations into polynomials with integer coefficients.

Lemma 2.3 (Gauss Lemma). *Let f be a polynomial with integer coefficients. Suppose $f = gh$ with $g, h \in \mathbb{Q}[X]$. Then there is an element $c \in \mathbb{Q}^\times$ such that both cg and $c^{-1}h$ have integer coefficients.*

In order to prove the lemma, we introduce the idea of a *primitive* polynomial in $\mathbb{Z}[X]$. By this, we mean a polynomial which is not a multiple of any prime number. For any non-zero polynomial $f \in \mathbb{Z}[X]$, there is a unique positive integer N such that $\frac{1}{N}f$ is primitive; in fact, N is the highest common factor of the coefficients of f. More generally for a non-zero polynomial $f \in \mathbb{Q}[X]$, there is a rational number $c > 0$ such that cf is primitive in $\mathbb{Z}[X]$. We shall first prove the following statement, which is equivalent to the Gauss Lemma.

Lemma 2.4. *If f and g are primitive polynomials in $\mathbb{Z}[X]$, then fg is also primitive.*

Proof. For any prime number p, we know that neither f nor g is a multiple of p. Therefore, in the ring $\mathbb{F}_p[X]$, we have $f \not\equiv 0 \bmod p$ and $g \not\equiv 0 \bmod p$. By Proposition 2.1, this implies $fg \not\equiv 0 \bmod p$. Therefore, fg is primitive. □

Proof of the Gauss Lemma. Choose non-zero rational numbers c and d such that cg and dh are primitive. Note that $cdf = (cg)(dh)$, so by

Lemma 2.4, we know that cdf is primitive. This implies $cd = \frac{1}{N}$, where N is the highest common factor of the coefficients of f. In particular, $c^{-1}h = Ndh \in \mathbb{Z}[X]$. □

In practice, we often use the following simple special case of the Gauss Lemma.

Corollary 2.4 (Monic Gauss Lemma). *Let f be a monic polynomial with integer coefficients. Suppose $f = gh$ with g and h monic polynomials in $\mathbb{Q}[X]$. Then g and h have coefficients in $\mathbb{Z}$. In particular, this means that f is irreducible over $\mathbb{Q}$ if and only if it is irreducible over $\mathbb{Z}$.*

Example. We'll use the Monic Gauss Lemma to show that the polynomial $f(X) = X^3 - 3X + 1$ is irreducible over $\mathbb{Q}$. Suppose for the sake of argument that f factorizes. By the Monic Gauss Lemma, there is a factorization of the following form:

$$f(X) = (X - a)(X^2 + bX + c), \quad a, b, c \in \mathbb{Z}.$$

Clearly, a is a root of f, so we would be finished if we could show that f has no roots in $\mathbb{Z}$. Expanding out the right-hand side of the factorization and comparing constant coefficients, we see that $ac = -1$. Since a and c are integers, this implies that $a = 1$ or $a = -1$. However, one can check that neither 1 nor -1 is a root of f. This gives a contradiction, so f is irreducible.

Generalizing the example above, we see that if f is a monic polynomial with integer coefficients, then every rational root of f is an integer and is also a factor of the constant term of f. By listing all the factors of the constant term, we can check whether f has a linear factor. This allows us to check rather easily whether a polynomial of degree 2 or 3 is irreducible over $\mathbb{Q}$. For polynomials of a degree bigger than 3, it becomes a little harder as the following example shows.

Example. We'll show that the polynomial $f(X) = X^4 + 4X^2 + 1$ is irreducible over $\mathbb{Q}$. By the argument above, the only possible rational roots are at 1 and -1 since 1 and -1 are the factors of the constant term 1. As $f(1) = f(-1) = 6 \neq 0$, it follows that f has no linear factor over $\mathbb{Q}$. Hence (using the Monic Gauss Lemma), any factorization would have the form

$$f(X) = (X^2 + aX + b)(X^2 + cX + d), \quad a, b, c, d \in \mathbb{Z}.$$

Expanding out the right-hand side of this equation and equating coefficients, we obtain the following simultaneous equations:

$$a + c = 0, \quad b + d + ac = 4,$$
$$ad + bc = 0, \qquad bd = 1.$$

The last of these equations implies $b = d = \pm 1$, and the first implies $c = -a$. Therefore, the second equation reduces to

$$\pm 2 - a^2 = 4.$$

This has no integer solutions, so in fact, there is no such factorization and f is irreducible. Obviously, the method used in this example can be adapted to deal with larger degree polynomials, but it becomes more difficult as the degree of the polynomial increases.

The reduction map. Another tool for showing that polynomials are irreducible over $\mathbb{Q}$ is to reduce modulo a prime number p. Suppose f is a polynomial with coefficients in $\mathbb{Z}$. By reducing the coefficients of f modulo p, we obtain a polynomial $\bar{f}$ with coefficients in $\mathbb{F}_p$. This process gives us a function

$$\mathbb{Z}[X] \to \mathbb{F}_p[X], \quad f \mapsto \bar{f}.$$

It's easy to check that $\overline{f+g} = \bar{f} + \bar{g}$ and $\overline{fg} = \bar{f}\bar{g}$. This amounts to saying that the function $f \mapsto \bar{f}$ is a ring homomorphism from $\mathbb{Z}[X]$ to $\mathbb{F}_p[X]$.

Corollary 2.5. *Let $f \in \mathbb{Z}[X]$ and assume that the leading coefficient of f is invertible modulo a prime number p. If $\bar{f}$ is irreducible over $\mathbb{F}_p$, then f is irreducible over $\mathbb{Q}$.*

Proof. Suppose $f = gh$ is a factorization in $\mathbb{Z}[X]$ where g and h have a smaller degree than f. Then $\bar{f} = \bar{g}\bar{h}$ in $\mathbb{F}_p[X]$. Since the leading coefficient of f is invertible modulo p, it follows that $\bar{f}$ has the same degree as f. Therefore, $\bar{g}$ and $\bar{h}$ have smaller degrees than $\bar{f}$, so we have a contradiction. □

Example. We'll show using Corollary 2.5 that the polynomial

$$f(X) = 17X^3 + 23X^2 + 181$$

is irreducible over $\mathbb{Q}$. To do this, we reduce modulo 2 and show that the polynomial $\bar{f}(X) = X^3 + X^2 + 1$ is irreducible over $\mathbb{F}_2$. This is easily seen by checking that $\bar{f}$ has no roots in $\mathbb{F}_2$:

$$\bar{f}(0) \equiv 1 \bmod 2, \quad \bar{f}(1) \equiv 1 \bmod 2.$$

In fact, Corollary 2.5 is not quite as useful as it appears. This is because it's possible for a polynomial to be irreducible over $\mathbb{Q}$, but still factorize modulo every prime number. An example of this is the polynomial $X^4 + 4X^2 + 1$, which we proved to be irreducible earlier. This factorizes modulo every prime (see Exercise 3.38). For example,

$$\begin{aligned} X^4 + 4X^2 + 1 &\equiv (X+1)^4 && \bmod 2, \\ &\equiv (X+1)^2(X+2)^2 && \bmod 3, \\ &\equiv (X^2+2X+4)(X^2-2X+4) && \bmod 5, \\ &\equiv (X^2+X-1)(X^2-X-1) && \bmod 7, \\ &\equiv (X+2)(X+5)(X+6)(X+9) && \bmod 11. \end{aligned}$$

Remark. For the benefit of a reader with some knowledge of Galois theory, we note that Corollary 2.5 may only be used to show that f is irreducible if the Galois group $\mathrm{Gal}(k/\mathbb{Q})$ of the splitting field k of f has an element of order $\deg(f)$. In the example above, the Galois group is a product of two cyclic groups of order 2, so has no element of order 4. It's for this reason that the polynomial factorizes modulo every prime, even though it is irreducible over $\mathbb{Q}$.

Another useful trick for showing that a polynomial is irreducible is the following.

Theorem 2.4 (Eisenstein's Criterion). *Let $f(X) = a_d X^d + \cdots + a_1 X + a_0$ be a polynomial with integer coefficients. Suppose there is a prime number p with the following properties:*

(1) *p is not a factor of the leading coefficient a_d;*
(2) *p is a factor of all the other coefficients $a_0, \ldots, a_{d-1}$;*
(3) *p^2 is not a factor of the constant term a_0.*

Then f is irreducible in $\mathbb{Q}[X]$.

Example. Consider the polynomial $f(X) = 5X^{18} + 27X^3 + 33$. The leading coefficient of f is 5, which is not a multiple of 3. The other coefficients are 27, 33 and several zeros, all of which are multiplies of 3. The constant term is 33, which is not a multiple of 9. Therefore, $f(X)$ is irreducible over $\mathbb{Q}$, as it satisfies Eisenstein's Criterion with the prime number 3. It would be hard to show this by the other methods that we've seen because f has large degree.

Proof of Eisenstein's Criterion. If f is not irreducible, then by the Gauss Lemma, there is a factorization $f = gh$, where g and h are in $\mathbb{Z}[X]$

and have smaller degree than f. We'll let r and s be the degrees of g and h, respectively. Our conditions on the coefficients imply that

$$\bar{g}(X)\bar{h}(X) \equiv \bar{f}(X) \equiv a_d X^d \mod p.$$

Here, we are writing $\bar{f}$ for the image of f in the ring $\mathbb{F}_p[X]$. By Uniqueness of factorization of polynomials in the ring $\mathbb{F}_p[X]$, it follows that

$$\bar{g}(X) \equiv cX^r \bmod p, \quad \bar{h}(X) \equiv dX^s \bmod p.$$

Therefore, the constant terms of g and h are multiplies of p. However, this implies that the constant term of f is a multiple of p^2, which contradicts the third assumption in the theorem. □

Eisenstein's Criterion gives a very quick way of showing that a polynomial of large degree is irreducible over $\mathbb{Q}$. However, this method cannot always be used. Most irreducible polynomials do not satisfy Eisenstein's criterion with any prime number p.

Exercise 2.11. Show that every polynomial over $\mathbb{R}$ with degree at least 3 is reducible.

Exercise 2.12.

(1) List all the monic polynomials over $\mathbb{F}_2$ of degree ≤ 3.
(2) Determine which of these polynomials are irreducible over $\mathbb{F}_2$.
(3) Factorize the reducible polynomials into irreducible polynomials.

Exercise 2.13. Prove that each of the following polynomials is irreducible over $\mathbb{Q}$:

$$X^{10} + 75, \quad X^3 + 5X^2 + 77, \quad X^5 + X^3 + X + 1.$$

Exercise 2.14. Factorize the polynomial $f(X) = X^4 - X^2 - 2 \in \mathbb{F}[X]$ into irreducibles when $\mathbb{F}$ is (i) $\mathbb{Q}$, (ii) $\mathbb{R}$, (iii) $\mathbb{C}$.

Exercise 2.15. Let f and g be polynomials over $\mathbb{Q}$. Show that f and g are coprime in $\mathbb{Q}[X]$ if and only if there is no complex number α such that $f(\alpha) = g(\alpha) = 0$.

2.5 UNIQUE FACTORIZATION DOMAINS

We have shown that in the rings $\mathbb{Z}$ and $\mathbb{F}[X]$, every non-zero element can be factorized uniquely into irreducible elements. In this section, we'll consider which other polynomial rings have this property. At this point, it's useful to use the following definition.

Definition. A ring R is called an *integral domain* if $1 \neq 0$ in R and for every $x, y \in R$ such that $xy = 0$, we must have either $x = 0$ or $y = 0$.

Clearly, every field is an integral domain, and it follows that every subring of a field is also an integral domain. In fact, the converse to this is also true: every integral domain is contained in a field (see Exercise 2.16). In an integral domain, if we have an equation of the form $px = py$ with p irreducible, then $x = y$. This cancellation property is important in proving that factorization is unique.

Definition. An integral domain R is called a *unique factorization domain* (often abbreviated UFD) if it satisfies the following two conditions:

(1) Every non-zero element of R is either a unit or a product of irreducible elements.
(2) If p is an irreducible element of R, and p is a factor of ab, then p is a factor of either a or b (or both).

The word "unique" in the phrase "unique factorization domain" needs some qualification. There are various ways to factorize ring elements, even in the ring $\mathbb{Z}$. For example, the number 6 factorizes as 2×3 and also as 3×2. So, we call two factorizations the same if we can simply write one of them in a different order to make them the same. Less trivially, we also have $2 \times 3 = (-2) \times (-3)$, and the numbers 2, 3, -2 and -3 are all irreducible in $\mathbb{Z}$. More generally, we may multiply the irreducible elements in a factorization by units in the ring to get new factorizations. This issue was not so apparent in $\mathbb{Z}$ or in $\mathbb{F}[X]$, since we were able to restrict our choice of irreducible elements to the prime numbers (which are all positive) or the monic polynomials. However, in an arbitrary UFD, there may be no obvious way of restricting the choice of irreducible factors to make their choice unique. Apart from these two caveats, the next proposition shows that factorization in a UFD is unique.

Proposition 2.6. *Let R is a unique factorization domain and suppose we have irreducible elements $p_1, \ldots, p_r, q_1, \ldots, q_s \in R$ satisfying*

$$p_1 \cdots p_r = q_1 \cdots q_s.$$

Then $r = s$ and we may reorder the elements $q_1, \ldots, q_s$ in such a way that each q_i has the form $u_i p_i$ for a unit $u_i \in R^\times$.

Proof. The proof is by induction on r. It is clearly true if $r = 1$. Assume the result for $r - 1$. Since p_1 is a factor of the product $q_1 \cdots q_s$, the second property of UFDs shows that p_1 is a factor of one of the elements q_i. After reordering $q_1, \ldots, q_s$, we may assume that p_1 is a factor of q_1, so $q_1 = u_1 p_1$ for some element $u_1 \in R$. However, q_1 is irreducible, so u_1 must be a unit. Since the ring R is an integral domain, we may cancel p_1 from both sides of the equation to get

$$u_1 p_2 \cdots p_r = q_2 \cdots q_s.$$

Note that $u_1 p_2$ is irreducible. Hence, by the inductive hypothesis, $r = s$ and we may reorder $q_2, \ldots, q_s$ in such a way that $q_2 = u_2 p_2, \ldots, q_r = u_r p_r$ for appropriate units u_i. □

So far, we've seen that $\mathbb{Z}$ is a UFD, and $\mathbb{F}[X]$ is a UFD for every field $\mathbb{F}$. The proof in each case was based on Euclid's algorithm, which allows us to divide by a non-zero element in such a way that the "remainder" is smaller than the element we divide by. If we try to generalize these proofs, then we are lead to the concept of a Euclidean domain, which is simply a ring in which one can use this method of proof.

Definition. An integral domain R is called a *Euclidean domain* if there is a function d from the non-zero elements of R to $\mathbb{N}$ with the following properties:

(1) For every $x, y \in R$ with $y \neq 0$, there exist $q, r \in R$ such that $a = qb + r$ and either $r = 0$ or $d(r) < d(y)$;
(2) If $x|y$, then $d(x) \leq d(y)$.

For example, the rings $\mathbb{Z}$ and $\mathbb{F}[X]$ are Euclidean domains. On $\mathbb{Z}$, we have the function $d(x) = |x|$, and on $\mathbb{F}[X]$, we have $d(f) = \deg(f)$. We'll see some more examples of Euclidean domains in Chapter 5.

Proposition 2.7. *Every Euclidean domain is a unique factorization domain.*

Sketch Proof. There are two properties to prove. The first of these is the existence of a factorization of every non-zero element $x \in R$ into irreducible elements. This is proved by induction on $d(x)$. For the second property, one simply modifies the proof of Euclid's Lemma for Polynomials. □

It turns out that there are many UFDs which are not Euclidean domains. The first example of such a ring that we come across is the ring $\mathbb{Z}[X]$. To see that $\mathbb{Z}[X]$ is not a Euclidean domain, we note that Bezout's Lemma does not hold in this ring. For example, the elements 2 and X have no common factor in $\mathbb{Z}[X]$ apart from ± 1. If there were any kind Euclidean algorithm in the ring $\mathbb{Z}[X]$, then we would be able to find polynomials $h, k \in \mathbb{Z}[X]$ satisfying $2 \cdot h(X) + X \cdot k(X) = 1$. However, such polynomials do not exist. Indeed, substituting $X = 0$ into this equation, we find that $h(0) = \frac{1}{2}$, and so h cannot have integer coefficients. This contradiction shows that $\mathbb{Z}[X]$ is not a Euclidean domain. On the other hand, we'll show that $\mathbb{Z}[X]$ is a UFD. The units in this ring are just the constant polynomials 1 and -1. There are two different kinds of irreducible element. First, note that each prime number p is irreducible when regarded as a constant polynomial in $\mathbb{Z}[X]$. Second, if p is an irreducible polynomial in $\mathbb{Q}[X]$, then there is a non-zero rational number c such that cp is a primitive element of $\mathbb{Z}[X]$, and it follows from the Gauss Lemma that cp is irreducible in $\mathbb{Z}[X]$. This gives us all the irreducible elements of $\mathbb{Z}[X]$.

Proposition 2.8. *The ring $\mathbb{Z}[X]$ is a unique factorization domain.*

Proof. The existence of a factorization into irreducibles is easy to see. It remains to show, that if p is an irreducible element of $\mathbb{Z}[X]$ and $p|fg$, then $p|f$ or $p|g$. Suppose first that p is a prime number. Then since $p|fg$, we know that $fg \equiv 0 \bmod p$. We can regard this congruence as an equation in the ring $\mathbb{F}_p[X]$. Since $\mathbb{F}_p$ is a field, Proposition 2.1 implies that either $f \equiv 0 \bmod p$ or $g \equiv 0 \bmod p$. Therefore, either $p|f$ or $p|g$. Next, consider the case that p is a primitive polynomial, which is irreducible over $\mathbb{Q}$. By uniqueness of factorization in $\mathbb{Q}[X]$, it follows that in the ring $\mathbb{Q}[X]$, either f or g is a multiple of p. Let's suppose for the sake of argument that $f = ph$ with $h \in \mathbb{Q}[X]$. By the Gauss Lemma, there is a non-zero rational number c such that both cp and $c^{-1}h$ are in $\mathbb{Z}[X]$. Since p is primitive, it follows that c is an integer, and therefore, h is in $\mathbb{Z}[X]$. Therefore, p is a factor of f in $\mathbb{Z}[X]$. □

In fact, one can generalize the proof of Proposition 2.8 to prove the following (see Exercise 2.18).

Proposition 2.9. *If R is a unique factorization domain, then $R[X]$ is a unique factorization domain.*

This has an interesting corollary that the rings $\mathbb{Z}[X_1, \ldots, X_n]$ and $\mathbb{F}[X_1, \ldots, X_n]$ of polynomials in several variables all have unique factorization. This can be proved by induction on n. To prove the inductive step, we note that

$$R[X_1, \ldots, X_{n+1}] = S[X_{n+1}], \quad \text{where } S = R[X_1, \ldots, X_n].$$

Exercise 2.16. This exercise shows that every integral domain is a subring of a field. Let R be an integral domain and let S be the set of pairs (a, b) with $a, b \in R$ and $b \neq 0$. We define a relation $\sim$ on the set S by $(a, b) \sim (a', b')$ if $ab' = ba'$.

(1) Show that $\sim$ is an equivalence relation on the set S.
(2) We'll write $\frac{a}{b}$ for the equivalence class of (a, b). Let $\mathbb{F}$ be the set of equivalence classes. Show that the following operations are well defined on $\mathbb{F}$:

$$\frac{a}{b} + \frac{c}{d} = \frac{ad + bc}{bd}, \qquad \frac{a}{b} \times \frac{c}{d} = \frac{ac}{bd}.$$

(3) Show that $\mathbb{F}$ is a field with the operations defined above and with elements $0 = \frac{0}{1}$ and $1 = \frac{1}{1}$.
(4) Show that R is isomorphic to the subring of $\mathbb{F}$ consisting of elements of the form $\frac{a}{1}$.
(5) Where in this question have you used the fact that R is an integral domain?

(The field $\mathbb{F}$ is called the *field of fractions of R*.)

Exercise 2.17. Show that if R is an integral domain, then $R[X]$ is an integral domain.

Exercise 2.18. Let R be a UFD. Prove a version of the Gauss Lemma for $R[X]$. Hence, prove Proposition 2.9.

HINTS FOR SOME EXERCISES

2.2. Use Corollary 2.3.

2.3. Use the Chinese Remainder Theorem for the first part of the question. You may also use Corollary 2.3 in the second part.

2.5. In the last part, substitute $X = \alpha$ into the equation $hf + kg = 1$.

2.11. Show that if $a + ib$ is a complex root of $f \in \mathbb{R}[X]$, then $X^2 - 2aX + (a^2 + b^2)$ is a factor of f.

2.15. Use Bezout's Lemma for Polynomials and the Fundamental Theorem of Algebra.

2.18. For the second part, modify the proof of Proposition 2.8.

Chapter 3

Congruences Modulo Prime Numbers

In this chapter, we continue with the theory of congruences in $\mathbb{Z}$, which we began in Chapter 1. The difference is that, in Chapter 1, we used no knowledge of mathematics beyond the integers, whereas in this chapter, we shall appeal to some other theories. We begin in Sections 3.1 and 3.2 by proving theorems of Fermat and Euler, which use ideas from the theory of finite groups. After that, we prove (from Section 3.3 onwards) theorems of Gauss, which require an understanding of polynomials.

Some ideas from group theory. Let G be a finite group. We'll write the group law in G as multiplication, and we'll write 1 for the identity element. As G is finite, the powers g^a of a fixed element $g \in G$ cannot all be different. Therefore, we can find integers a and b for which g^a and g^b are equal, even though $a \neq b$. This implies $g^n = 1$ where $n = a - b$. The smallest positive integer n such that $g^n = 1$ is called the *order* of g and is written $\mathrm{ord}(g)$. One can easily show that $g^a = g^b$ in G if and only if $a \equiv b \bmod \mathrm{ord}(g)$ (see Exercise 3.1).

We'll write $\langle g \rangle$ for the set of all powers of g in G, i.e.

$$\langle g \rangle = \{1, g, g^2, \ldots, g^{\mathrm{ord}(x)-1}\}.$$

The set $\langle g \rangle$ is closed under multiplication and contains the identity element 1. Therefore, $\langle g \rangle$ is a subgroup of G. Such subgroups are called *cyclic* subgroups. The group G itself is called a cyclic group if there is an element g such that $G = \langle g \rangle$. This is equivalent to saying that g has order $|G|$. The element g is called a *generator* of G.

We recall the following result from group theory (see, for example, [12]).

Theorem 3.1 (Lagrange's Theorem). *Let G be a finite group and let H be a subgroup of G. Then $|H|$ is a factor of $|G|$.*

We are actually most interested in the following corollary to Lagrange's Theorem.

Corollary 3.1. *Let G be a finite group. Then for every element $g \in G$, the order of g is a factor of $|G|$.*

Proof. By the discussion above, $\langle g \rangle$ is a subgroup with $\mathrm{ord}(x)$ elements. By Lagrange's Theorem, $\mathrm{ord}(x)$ is a factor of $|G|$. □

Exercise 3.1. Let g be an element of a finite group G. Show that $g^a = g^b$ if and only if $a \equiv b \bmod \mathrm{ord}(g)$.

Exercise 3.2. Find the order of each element of $\mathbb{F}_7^\times$. Verify Corollary 3.1 in this case and show that $\mathbb{F}_7^\times$ is cyclic.

To answer this in `sage`, we use `GF(7)` to define the field $\mathbb{F}_7$ and the command `multiplicative_order` to calculate the order of an element. For example, the following calculates the order of 6 in the group $\mathbb{F}_7^\times$:

```
F=GF(7)
F(6).multiplicative_order()
```

We can write a loop to list the orders of all the non-zero elements of $\mathbb{F}_7$ using the `for` command.

```
F = GF(7)
for x in F:
    if x != 0:                                (x!=0 means x ≠ 0).
        print(x,x.multiplicative_order())
```

Exercise 3.3. Let $\mathrm{GL}_2(\mathbb{F}_2)$ be the group of invertible 2×2 matrices with entries in the field $\mathbb{F}_2$. List the elements of $\mathrm{GL}_2(\mathbb{F}_2)$ and find the order of each element. Is this group cyclic?

This question can be answered in `sage` as follows:

```
G=GL(2,GF(2))
for g in G:
    pretty_print (g, ' has order ', g.order())
```

3.1 FERMAT'S LITTLE THEOREM

Theorem 3.2 (Fermat's Little Theorem). *Let p be a prime number and suppose x is an integer, which is not a multiple of p. Then we have* $x^{p-1} \equiv 1 \bmod p$.

Proof. Since $\mathbb{F}_p$ is a field, every element apart from 0 is invertible, so the set $\mathbb{F}_p^\times = \{1, 2, \ldots, p-1\}$ is a group with $p-1$ elements. Let d be the order of x, i.e. the smallest positive integer such that $x^d \equiv 1 \bmod p$. By Corollary 3.1, $p-1$ is a multiple of d. Therefore, x^{p-1} is a power of x^d, which is congruent to 1 modulo p. □

Fermat's little theorem simplifies many calculations. Suppose, for example, that we'd like to calculate powers of 10 modulo the prime number 19. By Fermat's Little Theorem, we have

$$10^{18} \equiv 1 \mod 19.$$

This shows that the congruency class $10^n \bmod 19$ depends only on the congruency class of n modulo 18. For example, since $182 \equiv 2 \bmod 18$, we have

$$10^{182} \equiv 10^2 \equiv 5 \mod 19.$$

Perhaps, the example above seems a little contrived because the power 182 is very close to a multiple of 18. More realistically, suppose we'd like to calculate 10^{102} modulo 19. Since $102 \equiv 12 \bmod 18$, we have

$$10^{102} \equiv 10^{12} \mod 19.$$

The power on the right-hand side is now quite small, and so we can do the rest of the calculation by hand. The first step is to note that $10^{12} = 100^6 \equiv 5^6 \bmod 19$. Therefore,

$$10^{102} \equiv 25^3 \equiv 6^3 \equiv 7 \mod 19.$$

Solving congruences with powers. We can also use Fermat's Little Theorem to solve congruences of the form

$$x^a \equiv b \mod p,$$

as long as p is a prime number and the power a is coprime to $p-1$.

If $b \equiv 0 \bmod p$, then since $\mathbb{F}_p$ is a field, the only solution is $x \equiv 0 \bmod p$. We may therefore assume that $b \not\equiv 0 \bmod p$. Hence, any solution x must also be non-zero modulo p. Fermat's Little Theorem then shows that the power $x^n \bmod p$ depends only on n modulo $p-1$.

To solve the congruence, we must find the inverse c of a modulo $p-1$. Raising both sides of the congruence to the power c, we have

$$x^{ac} \equiv b^c \mod p.$$

As $ac \equiv 1 \bmod p-1$, the left-hand side simplifies to x^1, so the solution is

$$x \equiv b^c \mod p.$$

As an example, we'll solve the congruence

$$x^5 \equiv 2 \mod 19.$$

Note that the power 5 is coprime to 18, so the method described above is applicable. The inverse of 5 modulo 18 is 11. Therefore, the solution is $x \equiv 2^{11} \bmod 19$. Of course, we might want to reduce $2^{11} = 2048$ modulo 19 to get a simpler answer. Dividing 2048 by 19, we get the remainder 15, and so we have

$$x \equiv 15 \mod 19.$$

Exercise 3.4. Calculate 3^{6478} modulo 79 and 3^{6478} modulo 83. Hence (using the Chinese Remainder Theorem), calculate 3^{6478} modulo 6557.

Exercise 3.5. Solve each of the following congruences:

$$x^5 \equiv 2 \bmod 7, \quad x^5 \equiv 3 \bmod 17, \quad x^7 \equiv 4 \bmod 111.$$

(Note that 111 is not prime, so the last congruence needs a slightly different method.)

The answers to the previous two questions can be checked on `sage`, for example, by typing

```
mod(3,79)^6478
     2
solve_mod(x^5==2, 7)
     4
```

Exercise 3.6. Let p and q be distinct odd prime numbers. Prove that for any integer x,

$$x^{\frac{pq-p-q+3}{2}} \equiv x \mod pq.$$

Exercise 3.7. Let $F : \mathbb{F}_p \to \mathbb{F}_p$ be any function. Show that there is a unique polynomial f of degree $\leq p-1$ such that $f(x) = F(x)$ for all $x \in \mathbb{F}_p$.

Exercise 3.8. (This question is another proof of Fermat's Little Theorem, so you should not use Fermat's Little Theorem in the answer.)

(1) Let p be a prime number and let a be an integer between 1 and $p-1$. Show that the binomial coefficient $\frac{p!}{a!(p-a)!}$ is a multiple of p.
(2) Hence, show that $(x+y)^p \equiv x^p + y^p \bmod p$ for all integers x and y.
(3) Hence, show, for all integers x, that $x^p \equiv x \bmod p$.

3.2 THE EULER TOTIENT FUNCTION

Recall that we write $(\mathbb{Z}/n)^\times$ for the set of invertible elements in $\mathbb{Z}/n$. These elements are the congruency classes of integers which are coprime to n. The set $(\mathbb{Z}/n)^\times$ is a group with the operation of multiplication modulo n. For example, the invertible congruency classes modulo 8 are 1, 3, 5 and 7, and multiplication in the group $(\mathbb{Z}/8)^\times = \{1, 3, 5, 7\}$ is given by the following table:

$\times$	1	3	5	7
1	1	3	5	7
3	3	1	7	5
5	5	7	1	3
7	7	5	3	1

Definition. We define $\varphi(n)$ to be the number of elements in the group $(\mathbb{Z}/n)^\times$. For example, $(\mathbb{Z}/8)^\times$ has 4 elements, so $\varphi(8) = 4$. We use the convention that $\varphi(1) = 1$; this makes sense because 0 is coprime to 1. The function φ is called the *Euler totient function.*[a]

Theorem 3.3 (Euler's Theorem). *For any* $x \in (\mathbb{Z}/n)^\times$, *we have* $x^{\varphi(n)} \equiv 1 \bmod n$.

[a]The word *totient* was invented by J. J. Sylvester in 1879 to describe the function, which had earlier been studied by Euler

Note that for a prime number p, all the integers $1, 2, \ldots, p-1$ are coprime to p, and so $\varphi(p) = p - 1$. Thus, when $n = p$, Euler's theorem is just a restatement of Fermat's Little Theorem.

Proof. This theorem is proved by the same method as Fermat's Little Theorem. Let d be the order of x. By Corollary 3.1, $\varphi(n)$ is a multiple of d. Since $x^d \equiv 1 \bmod n$, it follows that $x^{\varphi(n)} \equiv 1 \bmod n$. □

We can use Euler's theorem for calculations in $(\mathbb{Z}/n)^\times$ in the same way we used Fermat's Little Theorem in $\mathbb{F}_p^\times$. However, to be able to use the theorem, we first need to be able to calculate the numbers $\varphi(n)$. When n is small, one can simply count how many elements of $\mathbb{Z}/n$ are coprime to n. This method would be very slow for calculating a value such as $\varphi(1000)$. The next few results tell us how to calculate $\varphi(n)$ more quickly.

Calculating the Euler totient function. Suppose that n and m are coprime positive integers. According to the Chinese Remainder Theorem, there is a bijection $\mathcal{C} : \mathbb{Z}/nm \to \mathbb{Z}/n \times \mathbb{Z}/m$ defined by

$$\mathcal{C}(x) = (x \bmod n, x \bmod m).$$

With this notation, we have the following.

Lemma 3.1. *Let x be an element of $\mathbb{Z}/nm$. Then x is in $(\mathbb{Z}/nm)^\times$ if and only if $\mathcal{C}(x)$ is in $(\mathbb{Z}/n)^\times \times (\mathbb{Z}/m)^\times$.*

Proof. We must show that an integer x is coprime to nm if and only if x is coprime to both n and m. Suppose first that x is coprime to nm. Then clearly x has no common factor with either n or m.

Conversely, if x is coprime to both n and m, then no prime factor of either n or m is a factor of x. Hence (by Euclid's Lemma), no prime factor of nm is a factor of x, so x is coprime to nm. □

Proposition 3.1. *If n and m are coprime, then $\varphi(nm) = \varphi(n)\varphi(m)$.*

Proof. By the Lemma 3.1, the function $\mathcal{C}$ gives a bijection between the elements of $(\mathbb{Z}/nm)^\times$ and the elements of $(\mathbb{Z}/n)^\times \times (\mathbb{Z}/m)^\times$. The set $(\mathbb{Z}/nm)^\times$ has $\varphi(nm)$ elements and the product set $(\mathbb{Z}/n)^\times \times (\mathbb{Z}/m)^\times$ has $\varphi(n)\varphi(m)$ elements. □

In fact, the map $\mathcal{C} : (\mathbb{Z}/nm)^\times \to (\mathbb{Z}/n)^\times \times (\mathbb{Z}/m)^\times$ is more than just a bijection between two sets: it is a group isomorphism. Here, we regard the product set $(\mathbb{Z}/n)^\times \times (\mathbb{Z}/m)^\times$ as a group, where multiplication of pairs (x, y) is defined "componentwise" by the formula

$$(x, y) \times (x', y') = (xx', yy').$$

It is easy to check that this operation makes $(\mathbb{Z}/n)^\times \times (\mathbb{Z}/m)^\times$ into a group. More generally, for any groups G and H, the product set $G \times H$, is a group with the group operation defined by the formula above. The group $G \times H$ is called the *direct product* of the groups G and H.

Theorem 3.4. *If n and m are coprime, then $(\mathbb{Z}/nm)^\times$ is isomorphic to the direct product $(\mathbb{Z}/n)^\times \times (\mathbb{Z}/m)^\times$.*

Proof. We've already seen that the function $\mathcal{C} : (\mathbb{Z}/nm)^\times \to (\mathbb{Z}/n)^\times \times (\mathbb{Z}/m)^\times$ is a bijection. It remains to be shown that $\mathcal{C}(xy) = \mathcal{C}(x) \times \mathcal{C}(y)$. However, this follows straight from the definition:

$$\begin{aligned}\mathcal{C}(xy) &= (xy \bmod n, xy \bmod m)\\ &= (x \bmod n, x \bmod m) \times (y \bmod n, y \bmod m) = \mathcal{C}(x) \times \mathcal{C}(y).\end{aligned}$$ □

Lemma 3.2. *If p is prime and $a > 0$, then $\varphi(p^a) = (p-1)p^{a-1}$.*

Proof. By uniqueness of factorization, an integer has a common factor with p^a if and only if it is a multiple of p. Therefore, the elements of $\mathbb{Z}/p^a$ which are not invertible are $\{0, p, 2p, \ldots, p^a - p\}$. There are p^{a-1} of these, so the number of invertible elements is $p^a - p^{a-1} = (p-1)p^{a-1}$. □

Corollary 3.2. *Suppose $n = p_1^{a_1} \cdots p_r^{a_r}$ with p_i distinct primes and $a_i > 0$. Then*

$$\varphi(n) = (p_1 - 1)p_1^{a_1 - 1} \cdots (p_r - 1)p_r^{a_r - 1}.$$

Proof. The factors $p_1^{a_1}, \ldots, p_r^{a_r}$ are pairwise coprime, so by Proposition 3.1, we have $\varphi(n) = \varphi(p_1^{a_1}) \cdots \varphi(p_r^{a_r})$. The result follows from Lemma 3.2. □

Using Corollary 3.2, we can easily calculate values of φ. For example, to calculate $\varphi(1000)$, we factorize 1000 into primes and then use the formula in the corollary.

$$\begin{aligned}1000 &= 2^3 \cdot 5^3,\\ \varphi(1000) &= (2-1)2^{3-1} \cdot (5-1)5^{3-1} = 400.\end{aligned}$$

Solving congruences with powers. Euler's theorem allows us to do certain calculations very quickly. If x is coprime to n, then it tells us that $x^{\varphi(n)} \equiv 1 \bmod n$, and hence any power $x^a \bmod n$ depends only on the congruency class of a modulo $\varphi(n)$. For example, to calculate 2^{123} modulo 21, we first calculate $\varphi(21)$ and then reduce the power 123 modulo $\varphi(21)$. Note that $21 = 3 \times 7$, and so using Corollary 3.2, we have $\varphi(21) = 2 \times 6 = 12$. Since $123 \equiv 3 \bmod 12$, we have

$$2^{123} \equiv 2^3 \equiv 8 \mod 21.$$

We can also use Euler's theorem to solve congruences of the form

$$x^a \equiv b \mod n,$$

as long as a is invertible modulo $\varphi(n)$ and b is invertible modulo n. Note that any solution x to such a congruence must also be coprime to n. Hence, by Euler's Theorem, the power x^a depends only on the congruency class of a modulo $\varphi(n)$. Thus, to solve the congruence, we calculate the integer $c \equiv a^{-1} \bmod \varphi(n)$ and we have

$$x \equiv x^{ac} \equiv b^c \mod n.$$

As an example, we'll solve the congruence

$$x^7 \equiv 3 \mod 50. \tag{3.1}$$

In order to use the method described, we need to check that

- 3 is coprime to 50. This is certainly true.
- The power 7 is coprime to $\varphi(50)$. Since $50 = 2 \times 5^2$, we have $\varphi(50) = 1 \times 4 \times 5 = 20$. This is indeed coprime to 7.

To solve the congruence, we calculate the inverse of 7 modulo $\varphi(50) = 20$. This is done by Euclid's Algorithm as follows:

$$\begin{aligned} 20 &= 2 \times 7 + 6, & 1 &= 7 - 6 \\ 7 &= 6 + 1. & &= 7 - (20 - 2 \times 7) \\ & & &= 3 \times 7 - 20. \end{aligned}$$

Hence, $3 \times 7 \equiv 1 \bmod 20$, so the inverse of 7 modulo 20 is 3. Raising both sides of Equation (3.1) to the power 3, we get

$$x \equiv x^{7\times 3} \equiv 3^3 \equiv 27 \mod 50.$$

Therefore, the solution is $x \equiv 27 \bmod 50$.

Exercise 3.9. Calculate $\varphi(n)$ for $1 \leq n \leq 20$.

Exercise 3.10. Calculate $\varphi(10000)$. Hence, calculate $7^{135246872002}$ and $65^{123456789012345}$ modulo 10000 (note that 65 is not coprime to 10000).

Exercise 3.11. Solve each of the following congruences:

$$x^{461} \equiv 5 \bmod 2016, \quad x^{317} \equiv 3 \bmod 2345, \quad x^{1057} \equiv 2 \bmod 2467.$$

The questions above can be answered using `sage`. For example,

```
euler_phi(10000)
```

calculates $\varphi(10000)$. The command

```
solve_mod(x^461==5,2016)
```

solves the congruence $x^{461} \equiv 5 \bmod 2016$.

Exercise 3.12. Let B be a positive integer. Show that there are only finitely many positive integers n, such that $\varphi(n) \leq B$.

Exercise 3.13. This exercise shows that there are integers n such that $\frac{\varphi(n)}{n}$ is very small, so that a randomly chosen element $x \in \mathbb{Z}/n$ is almost certain to have a common factor with n.

(1) Show that for any prime p and any positive integer N,

$$\frac{p}{p-1} > 1 + \frac{1}{p} + \frac{1}{p^2} + \cdots + \frac{1}{p^N}.$$

(2) Hence, show that

$$\prod_{\text{primes } p<N} \left(\frac{p}{p-1}\right) > 1 + \frac{1}{2} + \frac{1}{3} + \cdots + \frac{1}{N}.$$

(3) Show that for every $\epsilon > 0$, there is a positive integer n such that $\frac{\varphi(n)}{n} < \epsilon$.

3.3 CYCLOTOMIC POLYNOMIALS AND PRIMITIVE ROOTS

Recall that a finite group G is called a *cyclic group* if there is an element $x \in G$ such that all the elements of G are powers of x. Equivalently, the order of x is $|G|$. For example, the additive group $(\mathbb{Z}/n, +)$ is cyclic of order n (the element 1 is a generator). In contrast, the multiplicative groups $(\mathbb{Z}/n)^\times$ are not always cyclic (see Exercises 3.17 and 3.18). In this section, we'll prove the following theorem of Gauss.

Theorem 3.5. *If p is a prime number, then $\mathbb{F}_p^\times$ is cyclic.*

In fact, we'll prove a more precise statement (see Theorem 3.6). Perhaps surprisingly, the theorem cannot be proved just using group theory, but requires some knowledge of the polynomial ring $\mathbb{F}_p[X]$. Before we get on to the proof, we'll describe some examples.

Definition. A generator of the cyclic group $\mathbb{F}_p^\times$ is called a *primitive root modulo p*. Note that x is a primitive root if and only if the order of x in $\mathbb{F}_p^\times$ is $p-1$, or equivalently if

$$\mathbb{F}_p^\times = \{1, x, x^2, \ldots, x^{p-2}\}.$$

Theorem 3.5 simply says that there is a primitive root modulo p.

Example. The powers of 3 in $\mathbb{F}_7$ are as follows:

$$\begin{array}{lll} 3^1 = 3, & 3^2 \equiv 2 \bmod 7, & 3^3 \equiv 6 \bmod 7, \\ 3^4 \equiv 4 \bmod 7, & 3^5 \equiv 5 \bmod 7, & 3^6 \equiv 1 \bmod 7. \end{array}$$

This shows that the element 3 has order 6 in $\mathbb{F}_7^\times$, so 3 is a primitive root modulo 7. A similar calculation shows that 5 is also a primitive root modulo 7. On the other hand, 2 is not a primitive root modulo 7. In fact, the element 2 has order 3 in $\mathbb{F}_7^\times$ because $2^3 \equiv 1 \bmod 7$.

To find out whether an integer x is a primitive root modulo p, we need to check whether x^n is congruent to 1 modulo p for any power $n < p-1$. At first sight, this looks like a big job. However, the following proposition means that we only need to test a few powers of x.

Proposition 3.2. *An element $x \in \mathbb{F}_p^\times$ is a primitive root modulo p if and only if $x^{\frac{p-1}{q}} \not\equiv 1 \bmod p$ for every prime factor q of $p-1$.*

Proof. Obviously, if $x^{\frac{p-1}{q}} \equiv 1 \bmod p$, then the order of x is no bigger than $\frac{p-1}{q}$, so x cannot be a primitive root.

Conversely, suppose x is not a primitive root, and let n be the order of x. By Corollary 3.1, we know that n is a proper factor of $p-1$, so $n = \frac{p-1}{d}$ for some $d > 1$. If q is a prime factor of d, then $\frac{p-1}{q}$ is a multiple of n. Therefore, $x^{\frac{p-1}{q}}$ is a power of x^n, which is congruent to 1. □

Example. We'll use Proposition 3.2 to find a primitive root modulo 29. The primes dividing $29 - 1 = 28$ are 2 and 7. As $\frac{28}{2} = 14$ and $\frac{28}{7} = 4$, the proposition gives us the following criterion:

x is a primitive root modulo 29 if and only if $x^{14} \not\equiv 1 \bmod 29$, and $x^4 \not\equiv 1 \bmod 29$.

For example, we can take $x = 2$. We have $2^4 = 16 \not\equiv 1 \bmod 29$ and $2^{14} = 128^2 \equiv 12^2 \equiv 144 \equiv 28 \bmod 29$. Therefore, 2 is a primitive root modulo 29.

When using the criterion in Proposition 3.2, we will always need to calculate the power $x^{\frac{p-1}{2}} \bmod p$. In fact, this is the largest power of x which we'll need to calculate. In Section 3.5, we'll see that there is a very fast method for calculating this power of x modulo p, since it is equal to the quadratic residue symbol $(\frac{x}{p})$.

Roots of unity and cyclotomic polynomials. To understand primitive roots in the field $\mathbb{F}_p$, we'll first look at roots of unity in the complex numbers. A complex number ζ is called an *n-th root of unity* if $\zeta^n = 1$. The n-th roots of unity have the form

$$\zeta = e^{2\pi i \frac{a}{n}}, \quad a = 0, 1, 2, \ldots, n-1.$$

We call ζ a *primitive* n-th root of unity if no smaller power of ζ is equal to 1, i.e. if n is the order of ζ in the group $\mathbb{C}^\times$. If ζ is not a primitive n-th root of unity, then it has some smaller order d than n, which means that $\frac{a}{n} = \frac{b}{d}$ for some integer b. Such a cancellation happens exactly when a and n have a common factor. For this reason, we see that the primitive n-th roots of unity are the complex numbers ζ for which $\mathrm{hcf}(a, n) = 1$, i.e. the elements

$$\zeta = e^{2\pi i \frac{a}{n}}, \quad \text{where } a \in (\mathbb{Z}/n)^\times.$$

In particular, there are exactly $\varphi(n)$ primitive n-th roots of unity, where φ is the Euler totient function.

For example, let $n = 4$. The primitive fourth roots of unity are i and i. These can be written as $e^{2\pi i(1/4)}$ and $e^{2\pi i(3/4)}$, and $(\mathbb{Z}/4)^\times = \{1, 3\}$.

The n-th *cyclotomic polynomial* $\Phi_n(X)$ is defined to be

$$\Phi_n(X) = \prod_{\substack{\text{primitive } n\text{-th} \\ \text{roots of unity } \zeta}} (X - \zeta).$$

The product is taken over all the primitive n-th roots of unity ζ in $\mathbb{C}$. Since there are $\varphi(n)$ primitive n-th roots of unity, the polynomial $\Phi_n(X)$ has degree $\varphi(n)$. The first few cyclotomic polynomials are expanded out in the following table (Table 3.1).

The next proposition will allow us to calculate the polynomials Φ_n quite easily.

Proposition 3.3. *For every positive integer n, we have $\prod_{d|n} \Phi_d(X) = X^n - 1$.*

Table 3.1: The first few cyclotomic polynomials.

n	primitive n-th roots of unity	$\Phi_n(X)$ definition	$\Phi_n(X)$ expansion
1	1	$X-1$	$X-1$
2	-1	$X-(-1)$	$X+1$
3	$e^{\frac{2\pi i}{3}},\ e^{\frac{4\pi i}{3}}$	$\left(X-e^{\frac{2\pi i}{3}}\right)\left(X-e^{\frac{4\pi i}{3}}\right)$	X^2+X+1
4	$i,\ -i$	$(X-i)(X-(-i))$	X^2+1

In this proposition, we are using the notation $\prod_{d|n}$ to mean that in the product, d runs over all the positive integers which are factors of n.

Proof. The roots of $X^n - 1$ are all the n-th roots of unity (whether primitive or not), so we have

$$X^n - 1 = \prod_{\substack{\text{all } n\text{-th} \\ \text{roots of unity } \zeta}} (X - \zeta).$$

Each n-th root of unity is a primitive d-th root of unity for exactly one of the factors d of n. We therefore have

$$X^n - 1 = \prod_{d|n} \left(\prod_{\substack{\text{primitive } d\text{-th} \\ \text{roots of unity } \zeta}} (X - \zeta) \right) = \prod_{d|n} \Phi_d(X).$$

□

In the case that n is a prime number p, Proposition 3.3 takes a very simple form. The factors of p are 1 and p, and so $X^p - 1 = \Phi_1(X)\Phi_p(X)$. This implies (using the formula for a geometric progression):

$$\Phi_p(X) = \frac{X^p - 1}{X - 1} = 1 + X + \cdots + X^{p-1}.$$

Even when n is not prime, we can calculate Φ_n recursively by first calculating Φ_d for the factors d of n. As an example, we'll calculate the polynomial Φ_{12}. Since $\varphi(12) = 4$, there are four primitive 12th roots of unity, and so Φ_{12} is a polynomial of degree 4. The formula of the proposition gives

$$\Phi_{12}(X) = \frac{X^{12} - 1}{\Phi_1(X)\Phi_2(X)\Phi_3(X)\Phi_4(X)\Phi_6(X)}.$$

We can see from Table 3.1 that $\Phi_4(X) = X^2+1$, and using Proposition 3.3 with $n = 6$, we get

$$\Phi_1(X)\Phi_2(X)\Phi_3(X)\Phi_6(X) = X^6 - 1.$$

Substituting this into our formula for Φ_{12}, we are left with

$$\Phi_{12}(X) = \frac{X^{12}-1}{(X^6-1)(X^2+1)}.$$

Since $X^{12} - 1 = (X^6+1)(X^6-1)$, we can cancel out the factor $X^6 - 1$ in the denominator. This gives us

$$\Phi_{12}(X) = \frac{X^6+1}{X^2+1} = X^4 - X^2 + 1.$$

Note that in all the examples we've seen, the polynomial $\Phi_n(X)$ has integer coefficients, even though it is defined as a polynomial over the complex numbers. The next proposition shows that this is always true.

Proposition 3.4. *The cyclotomic polynomials $\Phi_n(X)$ have coefficients in $\mathbb{Z}$.*

Proof. We'll prove the proposition by induction on n. It is certainly true when $n = 1$ because $\Phi_1(X) = X - 1$. Assume that $n > 1$ and that Φ_d has integer coefficients for all $d < n$.

Recall from Proposition 2.3 that if we have polynomials f and g over $\mathbb{Z}$ with g monic, then we may divide f by g with remainder. Both the quotient and the remainder will have coefficients in $\mathbb{Z}$. By Proposition 3.3, we have a factorization

$$X^n - 1 = f(X)\Phi_n(X), \quad \text{where } f(X) = \prod_{d|n,\ d<n} \Phi_d(X).$$

This means that Φ_n is the result of dividing the polynomial $X^n - 1$ by $f(X)$. It does not matter whether we divide in $\mathbb{C}[X]$ or $\mathbb{Q}[X]$ or $\mathbb{Z}[X]$; the process is the same. However, f is monic and by the inductive hypothesis we know that f has integer coefficients. Therefore, when we divide $X^n - 1$ by $f(X)$, the quotient and the remainder will have integer coefficients. In particular, Φ_n has integer coefficients. □

Existence of primitive roots. We're now ready to prove the existence of primitive roots modulo a prime number p. We'll actually prove the following slightly more precise statement.

Theorem 3.6. *For each factor n of $p-1$, there are exactly $\varphi(n)$ elements of order n in $\mathbb{F}_p^\times$. In particular, there are $\varphi(p-1)$ primitive roots.*

Proof. The idea is to show (just as in $\mathbb{C}$) that the elements of $\mathbb{F}_p^\times$ with order n are the roots of Φ_n. We'll show that each Φ_n has $\varphi(n)$ roots in $\mathbb{F}_p$, and so there are $\varphi(n)$ elements of order n.

We begin by factorizing the polynomial $f(X) = X^{p-1} - 1$ over the field $\mathbb{F}_p$. If we choose any element $a \in \mathbb{F}_p^\times$, then by Fermat's Little Theorem, we have

$$f(a) = a^{p-1} - 1 \equiv 0 \mod p.$$

This shows that a is a root of f, so by the Remainder Theorem $X - a$ must be a factor of $f(X)$ in $\mathbb{F}_p[X]$. The polynomial $f(X)$ has degree $p - 1$, and we have found $p - 1$ distinct linear factors $X - a$. It follows that we must have found all the factors,[b] and we have

$$f(X) \equiv \prod_{a \in \mathbb{F}_p^\times} (X - a) \mod p.$$

On the other hand, Proposition 3.3 gives a factorization of $f(X)$ into cyclotomic polynomials. Comparing these two factorizations, we get

$$\prod_{n \mid p-1} \Phi_n(X) \equiv \prod_{a \in \mathbb{F}_p^\times} (X - a) \mod p.$$

The factorization on the right-hand side is of the simplest possible kind: all the factors are linear and there are no repeated factors. From this, we conclude the following:

1. For each factor n of $p - 1$, the polynomial Φ_n is a product of distinct linear factors over $\mathbb{F}_p$. As Φ_n has degree $\varphi(n)$, it follows that Φ_n has exactly $\varphi(n)$ roots in $\mathbb{F}_p$.
2. Each element of $\mathbb{F}_p^\times$ is a root of Φ_n for exactly one of the factors n of $p - 1$.

To finish the proof, we'll show that the roots of Φ_n have order n in $\mathbb{F}_p^\times$.

Suppose $\Phi_n(a) \equiv 0 \bmod p$. We must show that $a^n \equiv 1 \bmod p$ and $a^m \not\equiv 1 \bmod p$ for any proper factor m of n. For the first equation, we use the factorization in Proposition 3.3:

$$a^n - 1 = \prod_{d \mid n} \Phi_d(a) \equiv 0 \mod p.$$

The product on the right-hand side is zero because it contains $\Phi_n(a)$. Now suppose $a^m \equiv 1 \bmod p$ for some proper factor m of n. This implies by

[b]Here, we are using the fact that factorization in $\mathbb{F}_p[X]$ is unique.

Proposition 3.3,

$$\prod_{d|m} \Phi_d(a) = a^m - 1 \equiv 0 \mod p.$$

Therefore, $\Phi_d(a) \equiv 0 \bmod p$ for some factor d of m. However, we are assuming that a is a root of Φ_n, and we've already seen that a cannot be a root of two of the cyclotomic polynomials, so we have a contradiction. □

Exercise 3.14. List all primitive roots modulo 5, 7 and 11.

Exercise 3.15. How many primitive roots are there modulo 29, 31, 37?

Exercise 3.16. Let n be coprime to 10 and let d be the order of 10 in the group $(\mathbb{Z}/n)^\times$. Show that the decimal expansion of $\frac{1}{n}$ is periodic with period length d. For example, $10^3 \equiv 1 \bmod 37$ and we have $\frac{1}{37} = 0.027027027\cdots$.

Exercise 3.17. List all elements of $(\mathbb{Z}/12)^\times$. Show that $(\mathbb{Z}/12)^\times$ is not a cyclic group.

Exercise 3.18. Let p and q be distinct odd prime numbers. Prove that for every element $x \in (\mathbb{Z}/pq)^\times$, $x^{\frac{\varphi(pq)}{2}} \equiv 1 \bmod pq$. Hence, show that $(\mathbb{Z}/pq)^\times$ is not cyclic.

Exercise 3.19. Let p be a prime number congruent to 1 modulo 4. Show that if g is a primitive root modulo p, then $-g$ is a primitive root modulo p.

Exercise 3.20. Let p be a prime number not equal to 3. Prove that the product of all the primitive roots modulo p is congruent to 1 modulo p.

Exercise 3.21. For every positive integer n, show that $\sum_{d|n} \varphi(d) = n$.

Exercise 3.22. Calculate the cyclotomic polynomials $\Phi_n(X)$ for $n = 6, 8, 10$. (This can be checked on `sage`, using the command `cyclotomic_polynomial(6)`.)

Exercise 3.23. Let n be odd. Show that ζ is a primitive n-th root of unity if and only if $-\zeta$ is a primitive $2n$-th root of unity. Hence, show that for odd $n \geq 3$, $\Phi_{2n}(X) = \Phi_n(-X)$.

Exercise 3.24. Let n be a positive integer and let p be a prime number.

(1) Assuming p is a factor of n, show that ζ is a primitive pn-th root of unity if and only if ζ^p is a primitive n-th root of unity. Hence, show that $\Phi_{pn}(X) = \Phi_n(X^p)$.

(2) Assuming p is not a factor of n, show that ζ^p is a primitive n-th root of unity if and only if either ζ is a primitive pn-th root of unity or ζ is a primitive n-th root of unity. Hence, show that $\Phi_{pn}(X) = \frac{\Phi_n(X^p)}{\Phi_n(X)}$.
(3) Show that $\Phi_{p^a}(1) = p$ for all prime powers $p^a > 1$. Show that $\Phi_n(1) = 1$ if $n > 1$ is not a power of a prime.

3.4 PUBLIC KEY CRYPTOGRAPHY

The theory described in the previous sections has important commercial applications in the field of cryptography. Before moving on, we'll discuss these applications briefly. A lot more information can be found in the book, *Elementary Number Theory: Primes, Congruences, and Secrets* [23].

It's traditional in the field of cryptography to suppose that there are two people called Alice and Bob, who would like to send secret messages to each other. Alice and Bob have never met; they communicate only over the internet, and we may assume that every message they send to each other will be intercepted and read by lots of other people. The problem is to find a way for Alice and Bob to communicate without anyone else being able to understand what they are saying. Of course, if Alice and Bob could agree on a secret code, then they could simply use this, and hope that nobody will break the code. However, since Alice and Bob can only communicate over the internet, it is rather difficult for them to agree on a secret code without the rest of the world finding out how to decode their messages. A coded message might be of the form

> "Dear Alice,
> Please send me product X. My credit card details are Y.
> Best wishes, Bob."

More usually in practice, the coded message is of the form "Dear Alice, let's use the following secret code to communicate ..." followed by a description of a secret code.

The solution to this problem is called "public key cryptography". This is a system in which Alice can tell Bob how to encode messages, but Bob (and the rest of the world) will not be able to deduce how to decode the encoded messages. The information on how to encode a message for Alice can be made public; this information is called Alice's "public key". Alice herself also keeps a "private key" which is the information required to decode her messages. She will not tell Bob her private key, because telling Bob this information will also tell the rest of the world how to read her messages. In this way, Bob is able to send coded messages to Alice, and only Alice will be able to decode them. It is crucial that one cannot easily work out the private key from the public key.

RSA cryptography. RSA is a method of public key cryptography, which is named after Rivest, Shamir and Adleman who first published the method in 1978 [19]. In the RSA system, Alice chooses two large prime numbers p and q and multiplies them together to form a very large number N. She also calculates the number $\varphi(N) = (p-1)(q-1)$. She chooses an integer a, which is coprime to $\varphi(N)$, and she calculates the integer $b = a^{-1} \bmod \varphi(N)$. The public key, which she can tell everyone, is the pair of numbers (N, a). She does not reveal the numbers p, q, $\varphi(N)$ or b, and she hopes that nobody can recover any of these numbers, given only knowledge of the public key (N, a).

When Bob wants to send a message to Alice, he represents the message as an element x in $\mathbb{Z}/N$. To encode the message, Bob calculates $c \equiv x^a \bmod N$. Then Bob sends just the coded message c.

When Alice receives the coded message c, she is able to recover the original message x by solving the congruence $x^a \equiv c \bmod N$. To do this, she uses the method described in Section 3.2. The solution is $x \equiv c^b \bmod N$. Alice can solve the congruence because she knows the number b.

Let's consider how difficult it is for anyone else to recover the message x, given just the numbers N, a and $x^a \bmod N$. In effect, we are asking whether there is a way of solving the congruence $x^a \equiv c \bmod N$. At first sight, this might seem easy, as we've discussed how to do it earlier in the chapter. One simply calculates $b = a^{-1} \bmod \varphi(N)$, and then we have $x \equiv c^b \bmod N$. However, in order to find b, it appears that we first need to calculate $\varphi(N) = (p-1)(q-1)$. If we could factorize N into the primes p and q, then we would immediately find $\varphi(N)$, and this would allow us to find b in the same way that Alice did. However, as p and q are very large prime numbers, it appears to be difficult to factorize N. One might ask whether there could be another way of recovering the number b without first factorizing N, but in fact any way of finding b would give us an easy way of factorizing N (see [19]).

There are some practical questions to consider, such as how difficult it is for Alice to come up with the numbers p, q and a in the first place. Let's suppose for the sake of argument that Alice wants the primes p and q to have size roughly e^{500}. She would also like them to be fairly random, so nobody can simply guess what they are. She can choose any integer s of size roughly e^{500} and then test all the integers s, $s+1$, $s+2$, etc. until she reaches a prime number, which she will call p. By the Prime Number Theorem, the probability of a random number of size roughly e^{500} being prime is roughly $\frac{1}{\log(e^{500})} = \frac{1}{500}$. She can therefore expect to test about 500 integers before she reaches her first prime number. The vast majority of these numbers can be ruled out straight away as they will be multiples of some fairly small primes. This leaves a few which have no small prime factor, and for which one needs to use a sophisticated method to test whether they are prime or not. The number of steps taken in an algorithm to test whether one of these numbers is prime is roughly $\log(e^{500})^6 = 500^6$. This gives Alice her first prime number p after far fewer than 500^7 steps. The second prime q can be found in a similar way, using a different random starting point instead of s. All of this is rather easy for a modern computer. In fact, the following `sage` command finds the first prime above e^{500}:

```
next_prime(floor(e^500)).
```

The command runs in about 2 seconds on a computer bought in 2008. The command `floor(e^500)` returns the largest integer $\leq e^{500}$.

The problem of finding a large integer a coprime to $\varphi(N)$ is easier than finding the actual primes p and q. In fact, a large proportion of the integers are coprime to $\varphi(N)$. To test whether any integer a is coprime to $\varphi(N)$, we use Euclid's algorithm with a and $\varphi(N)$. As a and $\varphi(N)$ have size roughly e^{1000}, Euclid's algorithm takes roughly $\log(e^{1000}) = 1000$ divisions with remainder. Each division with remainder takes time roughly 1000^2. Once a is chosen by Euclid's algorithm, we can calculate $b \equiv a^{-1} \bmod \varphi(N)$ by working through the steps of the algorithm backwards. This is much quicker than the problem of choosing the primes p and q.

To encode his message, Bob needs to calculate $x^a \bmod N$, and to decode the coded message, Alice needs to calculate $c^b \bmod N$. These calculations are of a similar kind. To perform the calculation quickly, Bob calculates, by repeated squaring modulo N, the congruency classes

$$x,\ x^2,\ x^4,\ x^8,\ \ldots, x^{2^r} \quad \text{where } 2^r \leq a < 2^{r+1}.$$

Each squaring operation takes roughly $\log(N)^2$ steps, and Bob does roughly $\log(N)$ of these squaring operations, taking time roughly $\log(N)^3$. He then multiplies roughly $\log(N)$ of these numbers together to obtain $x^a \bmod N$. Again, each of these multiplications takes roughly $\log(N)^2$ steps. Therefore, Bob takes roughly 1000^3 steps to encode his message. Alice needs to do a similar amount of work to decode the message and recover x from c.

The encryption seems to be secure as long as nobody is able to factorize an integer which is a product of two primes of size roughly e^{500}. Currently, the fastest known algorithm for factorizing an integer N into primes is much slower than the algorithm for testing whether a number is prime or not. This algorithm is the *number field sieve* (see [18]) and the time it takes to factorize N is roughly

$$\exp(d \cdot \log(N)^{\frac{1}{3}} \cdot \log\log(N)^{\frac{2}{3}}),$$

where the constant d is $\sqrt[3]{\frac{64}{9}}$, which is a little less than 2. In the case that N is roughly e^{1000}, this number is

$$\exp(d \cdot 1000^{\frac{1}{3}} \log(1000)^{\frac{2}{3}}) \approx \exp(d \cdot 10 \cdot \log(1000)^{\frac{2}{3}}) \approx e^{70}.$$

The crucial point to note here is that the time taken to come up with the public and private keys (which is roughly $500^7 \approx e^{44}$) is far smaller than the time taken to factorize N (roughly e^{70}). Alice hopes that her computer can cope with a calculation of size e^{44} in order to choose a key and that no computer in the world can cope with a calculation of size e^{70}. To some extent, this is wishful thinking. Indeed, there is currently no known proof that primality testing is significantly easier than factorizing. It's possible that a faster algorithm for factorizing N will be found and that this will be used to break the RSA codes. Nevertheless, RSA cryptography is widely used and trusted.

Diffie–Hellman key exchange. The Diffie–Hellman key exchange (invented by Diffie and Hellman in 1976 [8]) is a method for Alice and Bob to create a secret number k (called the *key*), which they both know but is known to nobody else, even though other people may read all the messages sent between Alice and Bob. The number k may then be used by both Alice and Bob to encode and decode messages.

To create the number k, Alice and Bob first choose large prime numbers p and q such that $p \equiv 1 \bmod q$. They also choose an element $g \in \mathbb{F}_p^\times$ with order q; such elements exist by Theorem 3.6. They send the numbers p and g to each other, so other people will be able to see these numbers as well. Next, Alice chooses a large integer a which is coprime to q. She then calculates $g^a \bmod p$, which is another element of order q. Alice sends the number g^a to Bob (she does not reveal the number a). Similarly, Bob chooses a large integer b coprime to q, and calculates the number $g^b \bmod p$, which he sends to Alice. Bob does not tell anyone what his number b is.

The number k, which Alice and Bob will use as their key, is the number

$$k \equiv g^{ab} \bmod p.$$

Alice can calculate this number without knowing b because Bob has told her the value of $g^b \bmod p$, and she can raise this number to the power a. Similarly, Bob knows the values of g^a and b, so he can also calculate $g^{ab} \bmod p$. Hence, both Alice and Bob can calculate the number k, but it seems rather hard for anyone else to calculate k. An outside observer will see the congruency classes g, g^a and g^b modulo p and will want to know the value of g^{ab} modulo p. If one could find the number a, given g and g^a modulo p, then one could simply calculate k in the same way that Alice does. However, there is no obvious fast way of recovering a from g and g^a. The problem of solving equations of the form $g^a \equiv h \bmod p$ is called the *discrete logarithm* problem and is thought to be hard if the order of g is a multiple of a large prime.

It is not difficult for Alice or Bob to come up with the numbers p, q and g. They first choose a large prime q. Next, they search for primes p of the form $nq + 1$ for small n. The first such prime will be reached after roughly $\log(q)$ attempts. To find an element g of order q in $\mathbb{F}_p^\times$, they choose a random element x and let $g = x^{\frac{p-1}{q}}$. By Fermat's Little Theorem, $g^q \equiv 1 \bmod p$, so either g has order q or $g \equiv 1 \bmod p$. In the unlikely event that $g \equiv 1 \bmod p$, they try again with a different x. In practice, this will never happen because the probability of randomly choosing an x such that $x^{\frac{p-1}{q}} \equiv 1 \bmod p$ is $\frac{1}{q}$.

It is rather important that we choose a number g whose order is a large prime q, or at least a multiple of a large prime. The reason is because if q factorizes as $q = rs$ with r fairly small, then the number $g_1 = g^s$ would have order r. To solve the discrete logarithm problem $g^x \equiv c \bmod p$, we could first solve $g_1^x \equiv c^s \bmod p$. Solving this second congruence would be easy, since we can list the powers of g_1; there are only a small number of them. Having solved the second congruence, we would know the congruency class of x modulo r, and we could then reduce our original problem to a discrete logarithm with an element $g_2 = g^r$ of order s. We'll see in the next chapter another method for calculating discrete logarithms in $(\mathbb{Z}/p^n)^\times$ where p is small but n is large.

As with RSA, there is a certain amount of wishful thinking. There seems to be no obvious fast algorithm for calculating discrete logarithms modulo large primes, but there is also no known proof that no such algorithm exists.

3.5 QUADRATIC RECIPROCITY

In this section, p will be an odd prime number, and we shall discuss quadratic equations modulo p. Recall that in Section 3.1, we considered congruences of the form $x^a \equiv b \bmod p$. We can often solve such a congruence by setting $x \equiv b^c \bmod p$, where c is the inverse of a modulo $p-1$. However, in the case $a = 2$, this method fails because 2 is not invertible modulo $p-1$. In fact, it's clear that such a simple method could not possibly work, since solutions do not always exist, and when they do exist, they are not unique (if $x^2 \equiv b \bmod p$ has a solution x, then it has another solution $-x$). It is actually quite hard to find solutions to quadratic congruences. Rather than solving quadratic congruences, we'll answer a slightly different question:

Which quadratic congruences have solutions in $\mathbb{F}_p$?

The quadratic residue symbol. Let p be an odd prime number and let $a \in \mathbb{F}_p^\times$. We'll say that a is a *quadratic residue* modulo p if the equation $x^2 \equiv a \bmod p$ has a solution $x \in \mathbb{F}_p^\times$. Otherwise, a is called a *quadratic non-residue*. We define the *quadratic residue symbol* $(\frac{a}{p})$ as follows:

$$\left(\frac{a}{p}\right) = \begin{cases} 1 & \text{if } a \text{ is a quadratic residue modulo } p, \\ -1 & \text{if } a \text{ is a quadratic non-residue modulo } p. \end{cases}$$

Example. Let $p = 7$. The squares in $\mathbb{F}_7^\times$ are listed in the following table.

x	1	2	3	4	5	6
$x^2 \bmod 7$	1	4	2	2	4	1

From the table, we see that 1, 2 and 4 are quadratic residues modulo 7, whereas 3, 5 and 6 are quadratic non-residues. Therefore, the quadratic residue symbol takes the following values:

a	1	2	3	4	5	6
$\left(\frac{a}{7}\right)$	1	1	-1	1	-1	-1

Note that we do not call 0 a quadratic residue or a quadratic non-residue, and we have not defined the symbol $\left(\frac{0}{p}\right)$. It is customary to define $\left(\frac{0}{p}\right) = 0$.

It turns out that there exist both quadratic residues and quadratic non-residues modulo p. For example, the squares $1^2 = 1$, $2^2 = 4$, etc. are quadratic residues. On the other hand, if g is a primitive root modulo p, then the next result shows that the odd powers of g are quadratic non-residues.

Proposition 3.5. *Let g be a primitive root modulo an odd prime p. For $r \in \mathbb{Z}$, the number g^r is a quadratic residue modulo p if and only if r is even.*

Proof. If r is even, then clearly g^r is a square, so it is a quadratic residue modulo p. Conversely, suppose that g^r is a quadratic residue. This means that g^r is the square of some element g^s, so we have $g^r \equiv g^{2s} \bmod p$. This implies $g^{r-2s} \equiv 1 \bmod p$. Since g has order $p-1$, it follows that the power $r-2s$ is a multiple of $p-1$. In particular, since $p-1$ is even, it follows that r must be even. □

As we stated earlier, p is an odd prime number in this section. It will be important to remember that this implies

$$\tfrac{p-1}{2} \textit{ is an integer.}$$

Corollary 3.3. *For any odd prime p, there are exactly $\frac{p-1}{2}$ quadratic residues and $\frac{p-1}{2}$ quadratic non-residues modulo p.*

Proof. The elements of $\mathbb{F}_p^\times$ are g^r with $r = 1, 2, \ldots, p-1$. Exactly half of these powers r are even, and the other half are odd. □

Theorem 3.7 (Euler's Criterion). *Let p be an odd prime and let $a \in \mathbb{F}_p^\times$. Then*

$$\left(\frac{a}{p}\right) \equiv a^{\frac{p-1}{2}} \mod p.$$

Proof. We must show that $a^{\frac{p-1}{2}}$ is congruent to 1 if a is a quadratic residue and -1 if a is a quadratic non-residue modulo p.

Let g be a primitive root modulo p and consider the number $h = g^{\frac{p-1}{2}}$. Since g has order $p-1$ in $\mathbb{F}_p^\times$, we know that $h^2 \equiv 1 \bmod p$, but $h \not\equiv 1 \bmod p$. Therefore, $h \equiv -1 \bmod p$.

If a is any element of $\mathbb{F}_p^\times$, then $a \equiv g^r$ for some integer r, and we have

$$a^{\frac{p-1}{2}} = h^r \equiv (-1)^r \mod p.$$

The right-hand side is 1 or -1 depending on whether r is even or odd, so the theorem follows from Proposition 3.5. □

Corollary 3.4. *Let p be an odd prime. For $a, b \in \mathbb{F}_p^\times$, we have*

$$\left(\frac{ab}{p}\right) = \left(\frac{a}{p}\right)\left(\frac{b}{p}\right).$$

Proof. This follows from Euler's Criterion since $(ab)^{\frac{p-1}{2}} = a^{\frac{p-1}{2}} b^{\frac{p-1}{2}}$. □

The Reciprocity Law. To find out whether a congruence $x^2 \equiv a \bmod p$ has solutions, we need to be able to calculate the quadratic residue symbol $\left(\frac{a}{p}\right)$. The following three theorems allow us to calculate this very quickly.

Theorem 3.8 (First Nebensatz). *If p is an odd prime, then $\left(\frac{-1}{p}\right) = (-1)^{\frac{p-1}{2}}$.*

Proof. Let p be an odd prime. By Euler's criterion, we have

$$\left(\frac{-1}{p}\right) \equiv (-1)^{\frac{p-1}{2}} \mod p.$$

Since both sides are ± 1 and $1 \not\equiv -1 \bmod p$, it follows that they are actually equal. □

The odd prime number p is congruent to either 1 or 3 modulo 4. If p is congruent to 1 modulo 4, then $\frac{p-1}{2}$ is even, and if p is congruent to 3 modulo 4, then $\frac{p-1}{2}$ is odd. We can therefore restate the First Nebensatz as follows:

$$\left(\frac{-1}{p}\right) = \begin{cases} 1 & \text{if } p \equiv 1 \bmod 4, \\ -1 & \text{if } p \equiv 3 \bmod 4. \end{cases}$$

Theorem 3.9 (Second Nebensatz). *If p is an odd prime, then $\left(\frac{2}{p}\right) = (-1)^{\frac{p^2-1}{8}}$.*

We can easily check that this is equivalent to

$$\left(\frac{2}{p}\right) = \begin{cases} 1 & \text{if } p \equiv 1 \text{ or } -1 \bmod 8, \\ -1 & \text{if } p \equiv 3 \text{ or } -3 \bmod 8. \end{cases}$$

Theorem 3.10 (Quadratic Reciprocity Law). *If p and q are distinct odd primes, then*

$$\left(\frac{p}{q}\right) = (-1)^{\frac{(p-1)(q-1)}{4}} \left(\frac{q}{p}\right).$$

The sign $(-1)^{\frac{(p-1)(q-1)}{4}}$ in the Quadratic Reciprocity Law depends on the parity of the integer $\frac{p-1}{2} \cdot \frac{q-1}{2}$. This integer is even if either $\frac{p-1}{2}$ or $\frac{q-1}{2}$ is even and is odd if $\frac{p-1}{2}$ and $\frac{q-1}{2}$ are both odd. Hence, for odd primes p and q, we have

$$(-1)^{\frac{p-1}{2} \cdot \frac{q-1}{2}} = \begin{cases} 1 & \text{if either } p \text{ or } q \text{ is congruent to 1 modulo 4,} \\ -1 & \text{if } p \text{ and } q \text{ are both congruent to 3 modulo 4.} \end{cases}$$

We can therefore restate the Quadratic Reciprocity Law as follows:

$$\left(\frac{p}{q}\right) = \begin{cases} \left(\frac{q}{p}\right) & \text{if either } p \text{ or } q \text{ is congruent to 1 modulo 4,} \\ -\left(\frac{q}{p}\right) & \text{if } p \text{ and } q \text{ are both congruent to 3 modulo 4.} \end{cases}$$

The word "Nebensatz" was used by Gauss, who proved these theorems, and means roughly "neighbouring theorem". For the moment, we shall just use the theorems in order to do calculations; the proofs of the theorems will be given in Sections 3.7 and 3.8.

Example. We'll find out whether the congruence $x^2 \equiv 37 \bmod 199$ has any solutions. Since 199 is prime, this is equivalent to calculating the quadratic residue symbol $\left(\frac{37}{199}\right)$. If the value is 1, then the congruence has solutions, and if it is -1, then there are no solutions.

The numbers 37 and 199 are distinct odd primes, and so by the Quadratic Reciprocity Law, we have

$$\left(\frac{37}{199}\right) = (-1)^{\frac{(37-1)(199-1)}{4}} \left(\frac{199}{37}\right).$$

The factor $(-1)^{\frac{(37-1)(199-1)}{4}}$ is 1 because 37 is congruent to 1 modulo 4. At first sight, the right-hand side appears just as difficult to calculate as the left-hand side because the numbers involved are just as big. However, $\left(\frac{199}{37}\right)$ depends only on the congruency class of 199 modulo 37. Since $199 \equiv 14 \bmod 37$, we have

$$\left(\frac{37}{199}\right) = \left(\frac{14}{37}\right).$$

The numerator 14 factorizes, so by Corollary 3.4, we have

$$\left(\frac{37}{199}\right) = \left(\frac{2}{37}\right)\left(\frac{7}{37}\right).$$

The Second Nebensatz shows that $\left(\frac{2}{37}\right) = -1$ because $37 \equiv -3 \bmod 8$. Substituting this, we have

$$\left(\frac{37}{199}\right) = -\left(\frac{7}{37}\right).$$

The numbers 37 and 7 are odd primes, so we may again use the Quadratic Reciprocity Law:

$$\left(\frac{37}{199}\right) = -\left(\frac{37}{7}\right).$$

The quadratic residue symbol on the right-hand side depends only on the congruency class of 37 modulo 7. Since $37 \equiv 2 \bmod 7$, this reduces to

$$\left(\frac{37}{199}\right) = -\left(\frac{2}{7}\right).$$

Using the Second Nebensatz again, we have

$$\left(\frac{37}{199}\right) = -1.$$

Therefore, the congruence $x^2 \equiv 37 \bmod 199$ has no solutions.

Example. Here is another congruence, which looks superficially more complicated:

$$x^2 - 3x - 5 \equiv 0 \mod 53.$$

We can reduce this to the previous kind of congruence by completing the square. This will always be possible modulo an odd prime because 2 is invertible modulo every odd prime.

$$\left(x - 2^{-1} \cdot 3\right)^2 \equiv 5 + (2^{-1} \cdot 3)^2 \mod 53.$$

Since $3 \equiv 56 \bmod 53$, we have $2^{-1} \cdot 3 \equiv 28 \bmod 53$, so we can rewrite the congruence as

$$y^2 \equiv 47 \mod 53, \quad \text{where } y = x - 28.$$

We can now see that the congruence has solutions if 47 is a quadratic residue modulo 53 and has no solutions if 47 is a quadratic non-residue. To find out whether the congruence has solutions, we calculate the quadratic residue symbol $\left(\frac{47}{53}\right)$ by the same method as in the previous example. We have

$$\begin{aligned}
\left(\frac{47}{53}\right) &= \left(\frac{53}{47}\right) && \text{by the Quadratic Reciprocity Law} \\
&= \left(\frac{6}{47}\right) && \text{since } 53 \equiv 6 \bmod 47 \\
&= \left(\frac{2}{47}\right)\left(\frac{3}{47}\right) && \text{by Corollary 3.4} \\
&= 1 \cdot \left(\frac{3}{47}\right) && \text{by the Second Nebensatz} \\
&= 1 \cdot (-1) \left(\frac{47}{3}\right) && \text{by the Quadratic Reciprocity Law} \\
&= -\left(\frac{2}{3}\right) && \text{since } 47 \equiv 2 \bmod 3 \\
&= 1 && \text{by the Second Nebensatz.}
\end{aligned}$$

It follows that 47 is a quadratic residue modulo 53. It's usually quite difficult to find the square root modulo p. However, in this case, note that $47 \equiv 100 = 10^2 \bmod 53$. Therefore, the solution to the original congruence is

$$x \equiv 28 \pm 10 \equiv 18 \text{ or } 38 \mod 53.$$

One can also turn the question around slightly: rather than asking which elements of $\mathbb{F}_p^\times$ are squares, we can fix an integer a, and ask for which primes p does the congruence $x^2 \equiv a \bmod p$ have solutions? For example, consider the following question:

For which primes p does the congruence $x^2 \equiv -6 \bmod p$ have solutions?

For most primes p, this question is answered by the quadratic residue symbol $\left(\frac{-6}{p}\right)$. However, $\left(\frac{a}{p}\right)$ is only defined when p is an odd prime and a is not a multiple of p. This means we can write down $\left(\frac{-6}{p}\right)$ as long as p is not equal to 2 or 3. We'll assume for a moment that p is a prime number not equal to 2 or 3. In this case, the congruence $x^2 \equiv -6 \bmod p$ has solutions if and only if $\left(\frac{-6}{p}\right) = 1$. We can calculate the quadratic residue symbol as follows:

$$\left(\frac{-6}{p}\right) = \left(\frac{-1}{p}\right)\left(\frac{2}{p}\right)\left(\frac{3}{p}\right) \quad \text{by Corollary 3.4.}$$

Using the First Nebensatz, Second Nebensatz and the Quadratic Reciprocity Law, we have

$$\left(\frac{-1}{p}\right) = (-1)^{\frac{p-1}{2}}, \quad \left(\frac{2}{p}\right) = (-1)^{\frac{p^2-1}{8}}, \quad \left(\frac{3}{p}\right) = (-1)^{\frac{(p-1)(3-1)}{4}}\left(\frac{p}{3}\right).$$

Substituting these, we get

$$\begin{aligned}\left(\frac{-6}{p}\right) &= (-1)^{\frac{p-1}{2}+\frac{p^2-1}{8}+\frac{(p-1)(3-1)}{4}}\left(\frac{p}{3}\right)\\ &= (-1)^{\frac{p^2-1}{8}}\left(\frac{p}{3}\right).\end{aligned}$$

The first factor $(-1)^{\frac{p^2-1}{8}}$ depends only on p modulo 8 and the second factor $\left(\frac{p}{3}\right)$ depends only on p modulo 3. Therefore, $\left(\frac{-6}{p}\right)$ is determined by the congruency class of p modulo 24. Since the prime p is not equal to 2 or 3, it is coprime to 24, so is congruent modulo 24, to one of the numbers $1, 5, 7, 11, 13, 17, 19, 23$. We'll calculate the quadratic residue symbol in each case.

$p \bmod 24$	$(-1)^{\frac{p^2-1}{8}}$	$\left(\frac{p}{3}\right)$	$\left(\frac{-6}{p}\right)$
1	1	1	1
5	−1	−1	1
7	1	1	1
11	−1	−1	1
13	−1	1	−1
17	1	−1	−1
19	−1	1	−1
23	1	−1	−1

Looking at the last column, we see that if p is congruent to 1, 5, 7 or 11 modulo 24, then the congruence $x^2 \equiv -6 \bmod p$ has solutions, whereas if p is congruent to 13, 17, 19 or 23, then there are no solutions. This leaves the primes 2 and 3 to consider. It's easy to see that -6 is a square modulo these primes because

$$\begin{aligned}-6 &\equiv 0^2 \mod 2,\\ &\equiv 0^2 \mod 3.\end{aligned}$$

Hence, -6 is a square modulo p if and only if $p = 2$ or $p = 3$ or $p \equiv 1, 5, 7$ or 11 modulo 24.

One might consider generalizing the question above, and asking, for any fixed polynomial $f \in \mathbb{Z}[X]$, the following question:

for which primes p does the congruence $f(x) \equiv 0 \bmod p$ have solutions?

Such questions often have no easily stated solution. However, if the complex roots of f are in the ring $\mathbb{Z}[e^{2\pi i/n}]$ for a suitable n, then the answer depends only on the congruency class of p modulo n (see, for example, Exercises 3.52 and 3.53 for examples of this phenomenon). The importance of the ring $\mathbb{Z}[e^{2\pi i/n}]$ will become clear in Sections 3.7 and 3.8. Questions of this kind lead to *class field theory*, which is beyond the scope of this book (see, for example, [5] or [4]).

Testing for primitive roots. Recall that to test whether an element $x \in \mathbb{F}_p^\times$ is a primitive root, we need to calculate the powers $x^{\frac{p-1}{q}} \bmod p$ for each prime factor q of $p-1$. If none of these powers are congruent to 1, then x is a primitive root. Unless $p = 2$, the prime number $q = 2$ is always a factor of $p-1$, and so we will always need to calculate $x^{\frac{p-1}{2}} \bmod p$. In fact, this is the largest power of x which we need to calculate. However, Euler's Criterion tells us that

$$x^{\frac{p-1}{2}} \equiv \left(\frac{x}{p}\right) \mod p.$$

We can therefore calculate $x^{\frac{p-1}{2}} \bmod p$ very quickly using the Reciprocity Law.

Example. For example, suppose we'd like to find a primitive root modulo 41. The prime factors of $41 - 1 = 40$ are 2 and 5. Hence, an element $x \in \mathbb{F}_{41}^\times$ is a primitive root modulo 41 if neither x^8 nor x^{20} is congruent to 1 modulo 41. As we described above, we can calculate x^{20} modulo 41 quickly using quadratic reciprocity.

- We first try $x = 2$. By Euler's Criterion, we have

$$2^{20} \equiv \left(\frac{2}{41}\right) \mod 41.$$

  Using the Second Nebensatz, we have $2^{20} \equiv 1 \bmod 41$. Therefore, 2 is not a primitive root modulo 41.
- Next, try $x = 3$. Again, by Euler's Criterion, we have

$$3^{20} \equiv \left(\frac{3}{41}\right) \mod 41.$$

  Using the Quadratic Reciprocity Law, we have

$$\left(\frac{3}{41}\right) = \left(\frac{41}{3}\right).$$

Since $41 \equiv 2 \bmod 3$, this reduces to

$$\left(\frac{3}{41}\right) = \left(\frac{2}{3}\right).$$

By the Second Nebensatz, we have $3^{20} \equiv -1 \bmod 41$. As this is not congruent to 1, we continue and check the power 3^8. A quick calculation shows that

$$3^8 \equiv 81^2 \equiv (-1)^2 \equiv 1 \mod 41.$$

Therefore, 3 is not a primitive root either.

- It's also clear that 4 is not a primitive root. There are two ways to see this. First, 4 is a perfect square, so is a quadratic residue modulo 41. This implies $4^{20} \equiv \left(\frac{4}{41}\right) \equiv 1 \bmod 41$. Secondly, we've already seen that the number 2 is not a primitive root. This means that not every element of $\mathbb{F}_{41}^{\times}$ is a power of 2. As every power of 4 is also a power of 2, it follows that not every element of $\mathbb{F}_{41}^{\times}$ is a power of 4. Hence, 4 is not a primitive root. This second argument is rather more general than the first; it shows that none of the numbers 8, 9, 16, 27, 32, etc. are primitive roots modulo 41 since we have already ruled out 2 and 3.
- We next check $x = 5$.

$$\begin{aligned} 5^{20} &\equiv \left(\frac{5}{41}\right) \mod 41 && \text{by Euler's Criterion} \\ &= \left(\frac{41}{5}\right) && \text{by the Quadratic Reciprocity Law} \\ &= \left(\frac{1}{5}\right) && \text{since } 41 = 1 \bmod 5 \\ &= 1 && \text{since 1 is a square modulo 5.} \end{aligned}$$

This shows that 5 is not a primitive root modulo 41.

- Next, try $x = 6$. We've already calculated the powers 2^{20}, 3^{20} and 3^8 modulo 41, so there is not much work left to calculate these powers of 6.

$$6^{20} \equiv 2^{20} \cdot 3^{20} \equiv 1 \cdot (-1) \equiv -1 \mod 41,$$

$$6^8 \equiv 2^8 \cdot 3^8 \equiv 256 \cdot 1 \equiv 10 \mod 41.$$

Since neither of these powers of 6 are congruent to 1, it follows that 6 is the first primitive root modulo 41.

Exercise 3.25. List the quadratic residues modulo 13 and for each of them give a square root.

Exercise 3.26. Calculate the following quadratic residue symbols:

$$\left(\frac{-2}{13}\right), \quad \left(\frac{96}{149}\right), \quad \left(\frac{-102}{199}\right), \quad \left(\frac{83}{181}\right).$$

One can check the answers on **sage**, for example, using the command

```
legendre_symbol(96,149)
```

to calculate $\left(\frac{96}{149}\right)$.

Exercise 3.27. Show that if $n \equiv \pm 1 \bmod 8$, then $(-1)^{\frac{n^2-1}{8}} = 1$, and if $n \equiv \pm 3 \bmod 8$, then $(-1)^{\frac{n^2-1}{8}} = -1$.

Exercise 3.28. For which prime numbers p does the congruence $x^2 \equiv 5 \bmod p$ have solutions?

Exercise 3.29. For which prime numbers p does the congruence $x^2 + x + 1 \equiv 0 \bmod p$ have solutions?

Exercise 3.30. Let p be a prime number congruent to 3 modulo 4. Prove that if a is a primitive root modulo p, then $-a$ is not a primitive root modulo p.

Exercise 3.31. Find the first primitive root modulo 97, 101, 151, 193. (Use quadratic reciprocity to speed up your calculation.)

Recall that one can check the answers on **sage** using the command

```
primitive_root(97)
```

Exercise 3.32. Show that if 2 is a primitive root modulo a prime p, then $p \equiv 3$ or $p \equiv 5 \bmod 8$.

Exercise 3.33. Let $N \geq 4$. Show that if p is a prime such that $p \equiv 1 \bmod (N!)$, then none of the numbers $1, 2, \ldots, N$ are primitive roots modulo p.

Exercise 3.34. Let p be a prime number which is congruent to 3 modulo 4. Prove that if $a \in \mathbb{F}_p^\times$ is a quadratic residue, then the congruence $x^4 \equiv a \bmod p$ has exactly two solutions in $\mathbb{F}_p$.

Exercise 3.35. Let p be a prime number which is congruent to 1 modulo 4. Prove that if $a \in \mathbb{F}_p^\times$ is a quadratic residue, then the congruence $x^4 \equiv a \bmod p$ has either no solutions or exactly four solutions in $\mathbb{F}_p$. Give examples of each case.

Exercise 3.36.

(1) Let n be an integer and let $N = 4n^2 + 1$. Show that every prime factor of N is congruent to 1 modulo 4.
(2) Prove that there are infinitely many prime numbers congruent to 1 modulo 4.

Exercise 3.37. By modifying the previous question, prove that there are infinitely many primes congruent to 1 modulo 6.

Exercise 3.38. Let $p \geq 5$ be a prime number. Show that at least one of the numbers -2, 3 or -6 is a quadratic residue modulo p. Hence, show that the polynomial $X^4 + 4X^2 + 1$ factorizes modulo p. (We saw in Section 2.4 that this polynomial is irreducible over $\mathbb{Q}$.)

Exercise 3.39. (**Generalized Reciprocity Law**) Suppose n is coprime to an odd positive integer m, where $m = p_1 \cdots p_r$ is the factorization of m into primes. Then we define the *Legendre symbol* by

$$\left(\frac{n}{m}\right) = \left(\frac{n}{p_1}\right) \cdots \left(\frac{n}{p_r}\right).$$

Prove that if n and m are both odd and coprime, then we still have

$$\left(\frac{n}{m}\right) = (-1)^{\frac{(n-1)(m-1)}{4}} \left(\frac{m}{n}\right).$$

Prove that if m is odd, then

$$\left(\frac{-1}{m}\right) = (-1)^{\frac{m-1}{2}}, \quad \left(\frac{2}{m}\right) = (-1)^{\frac{m^2-1}{8}}.$$

(Note: There is a significant advantage to using this more general Reciprocity Law if the numbers are large: it allows us to calculate $\left(\frac{a}{p}\right)$ without factorizing a into primes.)

Exercise 3.40.

(1) Let n be a positive integer and let $\chi : (\mathbb{Z}/n)^\times \to \{1, -1\}$ be a surjective group homomorphism. Show that there are infinitely many prime numbers p, such that $\chi(p) = -1$.
(2) Let n be a non-zero integer, and suppose that for all but finitely many prime numbers p, n is a quadratic residue modulo p. Prove that n is a perfect square.

3.6 CONGRUENCES IN AN ARBITRARY RING

Up until now, when we've discussed congruences they have been congruences in the ring $\mathbb{Z}$. However, we may study congruences in any ring. In the proofs of the Second Nebensatz and the Quadratic Reciprocity Law, we'll use congruences in certain rings called cyclotomic rings. It makes sense to point out some properties of congruences in general.

Let R be an arbitrary ring and let m be an element of R. We'll say that two elements $x, y \in R$ are *congruent modulo* mR if there is an element $z \in R$ such that $x - y = mz$. This is written as $x \equiv y \bmod mR$. The relation of congruence modulo mR is an equivalence relation on R. The equivalence classes are called *congruency classes.* The congruency class of an element $x \in R$ is the set of elements $y \in R$, which are congruent to x modulo mR, i.e. the set

$$\{x + mz : z \in R\}.$$

The set of all congruency classes is written as R/mR. With this notation, $x \equiv y \bmod mR$ is equivalent to saying that the congruency class of x is equal to the congruency class of y in R/mR.

We would like to make R/mR into a ring, in the same way that we made $\mathbb{Z}/n$ into a ring in Chapter 1. In other words, we would like to be able to add and multiply congruency classes in R/mR in a way which is compatible with addition and multiplication in R. Given two congruency classes ($x \bmod mR$) and ($y \bmod mR$), we define their sum and product to be ($x + y \bmod mR$) and ($xy \bmod mR$). To check that this makes sense, we need to be sure that the congruency classes of xy and $x + y$ are not changed if we replace x and y by elements in the same congruency classes as x and y. This is guaranteed by the next lemma.

Lemma 3.3. *Let $x, x', y, y' \in R$. If $x \equiv x' \bmod mR$ and $y \equiv y' \bmod mR$, then $x + y \equiv x' + y' \bmod mR$ and $xy \equiv x'y' \bmod mR$.*

Proof. We know that $x - x' = ma$ and $y - y' = mb$ for elements a and b in R. Therefore, $(x + y) - (x' + y') = m(a + b)$ and $xy - x'y' = m(ay + x'b)$. □

The lemma implies that the operations of addition and multiplication on R/mR are well defined. To show that R/mR is a ring, we must also check that these operations satisfy the ring axioms. However, as with $\mathbb{Z}/n$, it's easy to deduce each ring axiom for R/mR from the corresponding ring axiom for R. For example, one of the ring axioms in R/mR is the congruence

$$x + y \equiv y + x \mod mR.$$

This is an obvious consequence of the equation in R:

$$x + y = y + x,$$

which we know to be true since R itself satisfies the ring axioms. The ring R/mR which we've just defined is called a *quotient ring*.

Congruences between polynomials. When dealing with the quotient ring $\mathbb{Z}/n$, we have a convenient set of integers $S = \{0, 1, \ldots, n-1\}$ which contains one representative of each congruency class. Given any integer x, we can find the representative $r \in S$ by dividing x by n and taking the remainder:

$$x = qn + r, \quad r \in S.$$

One can then regard $\mathbb{Z}/n$ as the set S on the understanding that in order to add or multiply in $\mathbb{Z}/n$, we first add or multiply the integers and then reduce the answer modulo n.

A similar thing happens in quotients of $\mathbb{F}[X]$, where $\mathbb{F}$ is a field. Let m be a non-zero polynomial over a field $\mathbb{F}$ with degree d. For any polynomial f, we may divide f by m with remainder:

$$f = qm + r, \quad \deg r < d.$$

This formula implies that $f \equiv r \bmod m\mathbb{F}[X]$. If we let S be the set of polynomials of degree less than d, then this shows that S contains a representative from each congruency class modulo $m\mathbb{F}[X]$. Furthermore, no two elements of S are congruent modulo $m\mathbb{F}[X]$, since every non-zero multiple of m has degree at least d. Therefore, the set S plays the same role for $\mathbb{F}[X]/m\mathbb{F}[X]$ as $\{0, \ldots, n-1\}$ does in $\mathbb{Z}/n$. We may regard $\mathbb{F}[X]/m\mathbb{F}[X]$ as the set S, on the understanding that after multiplying two polynomials, we must replace the answer by its remainder after dividing by m. Abusing this notation slightly, we generally identify $\mathbb{F}[X]/m\mathbb{F}[X]$ with the set S and write

$$\mathbb{F}[X]/m\mathbb{F}[X] = \{a_0 + a_1 X + \cdots + a_{d-1} X^{d-1} : a_0, \ldots, a_{d-1} \in \mathbb{F}\}, \quad d = \deg m.$$

As an example, we'll consider the polynomial $m(X) = X^2 + X + 1$ over the field $\mathbb{F}_2$. The polynomial m has degree 2, and so the set of representatives consists of polynomials $aX + b$, where a and b are in $\mathbb{F}_2$. This means that $\mathbb{F}_2[X]/m\mathbb{F}_2[X]$ has four elements:

$$\mathbb{F}_2[X]/m\mathbb{F}_2[X] = \{0, 1, X, X+1\}.$$

Addition in $\mathbb{F}_2[X]/m\mathbb{F}_2[X]$ is just the same as addition of polynomials over $\mathbb{F}_2$. In order to multiply elements, we need to be able to cancel an X^2 term. Note that since $X^2 + X + 1 \equiv 0 \bmod m\mathbb{F}_2[X]$, we have

$$X^2 \equiv -X - 1 \equiv X + 1 \mod m\mathbb{F}_2[X].$$

Using this fact, we can easily multiply any two elements of the quotient ring: we simply multiply in the usual way in $\mathbb{F}_2[X]$ and then replace any X^2 term by $X + 1$. Doing this, we obtain the multiplication table for the non-zero elements.

$\times$	1	X	$X+1$
1	1	X	$X+1$
X	X	$X+1$	1
$X+1$	$X+1$	1	X

We can see from the table that every element of this quotient ring has an inverse, and so $\mathbb{F}_2[X]/m\mathbb{F}_2[X]$ is a field. The next two results explain why this is the case; the statements and proofs are similar to Proposition 1.2 and Theorem 1.2, which are the corresponding results for $\mathbb{Z}/n$.

We say that f is invertible modulo $m\mathbb{F}[X]$ if there is a polynomial g such that $fg \equiv 1 \bmod m\mathbb{F}[X]$. This is equivalent to saying that the congruency class of f is invertible in the quotient ring $\mathbb{F}[X]/m\mathbb{F}[X]$.

Proposition 3.6. *A polynomial f is invertible modulo $m\mathbb{F}[X]$ if and only if f and m are coprime in $\mathbb{F}[X]$.*

Proof. Assume f is invertible modulo $m\mathbb{F}[X]$ and let g be its inverse. The congruence $fg \equiv 1 \bmod m\mathbb{F}[X]$ means that $fg = 1 + mh$ for some polynomial h. From this equation, we see that any common factor of f and m must also be a factor of 1. Therefore, f and m are coprime.

Conversely, if we assume that f and m are coprime, then we can find polynomials h, k such that $hf + km = 1$. This implies $hf \equiv 1 \bmod m\mathbb{F}[X]$. □

Corollary 3.5. *If m is an irreducible polynomial over a field $\mathbb{F}$, then $\mathbb{F}[X]/m\mathbb{F}[X]$ is a field.*

Proof. Since m is not constant, it follows that $1 \not\equiv 0 \bmod m\mathbb{F}[X]$. If $f \not\equiv 0 \bmod m\mathbb{F}[X]$, then since m is irreducible, it follows that f and m are coprime. Therefore, f is invertible modulo $m\mathbb{F}[X]$. □

In particular, thc corollary shows that since $X^2 + X + 1$ is irreducible over $\mathbb{F}_2$, the quotient ring in the example above is a field.

Algebraic integers. As another example, let's suppose m is a monic polynomial over $\mathbb{Z}$ of degree d. Since m is monic, we may divide by m with remainder in the ring $\mathbb{Z}[X]$. As before, this shows that every polynomial $f \in \mathbb{Z}[X]$ is congruent modulo $m\mathbb{Z}[X]$ to a unique polynomial r of degree less than d. The polynomial r is the remainder which we obtain when dividing f by m. This shows that

$$\mathbb{Z}[X]/m\mathbb{Z}[X] = \{a_0 + a_1 X + \cdots + a_{d-1}X^{d-1} : a_0, \ldots, a_{d-1} \in \mathbb{Z}\}.$$

To multiply in $\mathbb{Z}[X]/m\mathbb{Z}[X]$, we first multiply the elements as polynomials and then replace the answcr by its remainder when divided by m. In this sense, quotients of $\mathbb{Z}[X]$ by monic polynomials look quite similar to quotients of $\mathbb{F}[X]$. We'll give another interpretation of the quotients rings $\mathbb{Z}[X]/m\mathbb{Z}[X]$ in the case that m is monic and irreducible.

A complex number α is called an *algebraic integer* if α is a root of some monic polynomial f with integer coefficients. Note that if α is a root of f, then α must be a root of one of the irreducible factors m of f. We shall therefore assume that α is a root of an irreducible monic polynomial $m \in \mathbb{Z}[X]$. For example, $\sqrt{2}$ is an algebraic integer; it is a root of the irreducible monic polynomial $X^2 - 2$.

Given an algebraic integer α, we'll use the notation

$$\mathbb{Z}[\alpha] = \{f(\alpha) : f \in \mathbb{Z}[X]\}.$$

This means that the elements of $\mathbb{Z}[\alpha]$ are finite sums of the form

$$a_0 + a_1\alpha + a_2\alpha^2 + \cdots + a_n\alpha^n, \quad \text{where } a_0, \ldots, a_n \in \mathbb{Z}.$$

It is clear that $\mathbb{Z}[\alpha]$ is a subring of $\mathbb{C}$, since it contains 0 and 1 and is closed under the operations of addition, subtraction and multiplication (see Exercise 1.16).

An element of $\mathbb{Z}[\alpha]$ may be expressed as $f(\alpha)$ for many different polynomials f. For example, $(\sqrt{2})^4 + 1$ is the same number as $(\sqrt{2})^6 - 3$, and both of these numbers are in fact 5. It's useful to be able to write elements of $\mathbb{Z}[\alpha]$ in a form which is unique, as this allows us to see straightaway whether two elements are equal. For this, we use the following.

Lemma 3.4. *Let $\alpha \in \mathbb{C}$ be an algebraic integer, which is a root of an irreducible monic polynomial $m \in \mathbb{Z}[X]$. For polynomials f and g in $\mathbb{Z}[X]$, we have $f(\alpha) = g(\alpha)$ if and only if $f \equiv g \bmod m\mathbb{Z}[X]$.*

Proof. If $f \equiv g \bmod m\mathbb{Z}[X]$, then there is a polynomial h such that $f(X) = g(X) + m(X)h(X)$. Substituting $X = \alpha$ and using the fact that $m(\alpha) = 0$, we get $f(\alpha) = g(\alpha)$.

Conversely, let's assume that $f(\alpha) = g(\alpha)$ for two polynomials $f, g \in \mathbb{Z}[X]$. Since m is a monic polynomial, we may divide $f - g$ by m with remainder in $\mathbb{Z}[X]$:

$$f(X) - g(X) = q(X)m(X) + r(X), \quad \deg r < \deg m.$$

To show that $f \equiv g \bmod m\mathbb{Z}[X]$, we must show that r is the zero polynomial. For the sake of argument, we'll assume that $r \neq 0$ and derive a contradiction. Substituting $X = \alpha$ in the equation above, we find that $r(\alpha) = 0$. In the larger ring $\mathbb{Q}[X]$, the polynomial m is still irreducible by the Monic Gauss Lemma. Since r has a smaller degree than m, we know that m cannot be a factor of r, and so r and m are coprime in $\mathbb{Q}[X]$. We can therefore find polynomials $h, k \in \mathbb{Q}[X]$ such that

$$h(X)r(X) + k(X)m(X) = 1.$$

Substituting α for X into this equation, and using the fact that $r(\alpha) = m(\alpha) = 0$, we see that it implies $0 = 1$. We therefore have a contradiction. □

Corollary 3.6. *Let α be a root of an irreducible monic polynomial $m \in \mathbb{Z}[X]$ of degree d. Then every element of $\mathbb{Z}[\alpha]$ can be written uniquely in the form*

$$a_0 + a_1\alpha + a_2\alpha^2 + \cdots + a_{d-1}\alpha^{d-1}, \quad a_0, a_1, a_2, \ldots, a_{d-1} \in \mathbb{Z}.$$

Proof. This follows from Lemma 3.4, since every congruency class in $\mathbb{Z}[X]/m\mathbb{Z}[X]$ has a unique representative of the form $a_0 + a_1X + a_2X^2 + \cdots + a_{d-1}X^{d-1}$. □

Congruences modulo a prime number. If p is a prime number, then we may regard p as an element of any ring R by letting $p = 1 + 1 + \cdots + 1$, where we add the element $1 \in R$ to itself p times. Of course, there are rings in which p is the zero element (for example, p is the zero element in $\mathbb{F}_p$). Congruences modulo pR have the following property, which we shall use in the proof of the Reciprocity Laws.

Lemma 3.5. *Let p be a prime number and let R be an arbitrary ring. Then for all elements $x, y \in R$ we have*

$$(x+y)^p \equiv x^p + y^p \mod pR.$$

Proof. By Exercise 3.8, we know that all but the first and last binomial coefficients $\frac{p!}{a!(p-a)!}$ are multiples of p. Hence, the formula is just the binomial expansion. □

Now, let α be an algebraic integer, which is a root of an irreducible monic polynomial $m \in \mathbb{Z}[X]$ of degree d. We've seen that elements of $\mathbb{Z}[\alpha]$ may be written uniquely in the form

$$a_0 + a_1\alpha + \cdots + a_{d-1}\alpha^{d-1},$$

with coefficients $a_0, \ldots, a_{d-1} \in \mathbb{Z}$. For elements $A, B \in \mathbb{Z}[\alpha]$, the congruence

$$A \equiv B \mod p\mathbb{Z}[\alpha]$$

means that $A - B = pC$ for some element C in $\mathbb{Z}[\alpha]$. If $A = \sum_{r=0}^{d-1} a_r\alpha^r$ and $B = \sum_{r=0}^{d-1} b_r\alpha^r$, then the congruence $A \equiv B \bmod p\mathbb{Z}[\alpha]$ is equivalent to the following set of simultaneous congruences in $\mathbb{Z}$ between the coefficients of A and B:

$$a_0 \equiv b_0 \bmod p, \quad a_1 \equiv b_1 \bmod p, \quad \ldots \quad a_{d-1} \equiv b_{d-1} \bmod p.$$

This means that every element of $\mathbb{Z}[\alpha]/p\mathbb{Z}[\alpha]$ is congruent to a unique element of the form

$$a_0 + a_1\alpha + \cdots + a_{d-1}\alpha^{d-1}, \quad a_i \in \{0, 1, \ldots, p-1\}.$$

In particular, the quotient ring $\mathbb{Z}[\alpha]/p\mathbb{Z}[\alpha]$ is finite with p^d elements. Even though p is a prime number, we do not usually expect $\mathbb{Z}[\alpha]/p\mathbb{Z}[\alpha]$ to be a field. Indeed, $\mathbb{Z}[\alpha]/p\mathbb{Z}[\alpha]$ is isomorphic to $\mathbb{F}_p[X]/m\mathbb{F}_p[X]$, and so $\mathbb{Z}[\alpha]/p\mathbb{Z}[\alpha]$ would only be a field if m were irreducible over $\mathbb{F}_p$.

Lemma 3.6. *Let α be an algebraic integer and let n be a positive integer. For $A, B \in \mathbb{Z}$, we have $A \equiv B \bmod p\mathbb{Z}[\alpha]$ if and only if $A \equiv B \bmod p\mathbb{Z}$.*

Proof. If our elements A and B are in $\mathbb{Z}$, then only the coefficients a_0 and b_0 are non-zero, and so the congruence $A \equiv B \bmod p\mathbb{Z}[\alpha]$ is equivalent to the usual congruence $A \equiv B \bmod p$. □

In particular, for an odd prime number p, we have

$$1 \not\equiv -1 \mod p\mathbb{Z}[\alpha]. \tag{3.2}$$

This shows that in order to calculate a quadratic residue symbol $\left(\frac{a}{p}\right)$, it is sufficient to calculate its congruency class modulo $p\mathbb{Z}[\alpha]$ in one of the rings $\mathbb{Z}[\alpha]$.

Exercise 3.41. Let R be any ring and let m be an element of R. Show that congruence modulo mR is an equivalence relation.

Exercise 3.42. (Chinese Remainder Theorem for polynomial rings) Let f and g be polynomials over a field $\mathbb{F}$ and assume that f and g are coprime in $\mathbb{F}[X]$. Show that for every $a, b \in \mathbb{F}[X]$, there is a unique $c \in \mathbb{F}[X]/fg\mathbb{F}[X]$ such that $c \equiv a \bmod f\mathbb{F}[X]$ and $c \equiv b \bmod g\mathbb{F}[X]$.

Exercise 3.43. For every prime number p, show that there is a field with p^2 elements.

Exercise 3.44. Show that 5, $\sqrt[3]{6}$ and $\sqrt{2} + \sqrt{3}$ are algebraic integers.

Exercise 3.45. Let α be a root of an irreducible monic polynomial $m \in \mathbb{Z}[X]$. Show that the map $\alpha^* : \mathbb{Z}[X]/m\mathbb{Z}[X] \to \mathbb{Z}[\alpha]$ defined by $\alpha^*(f) = f(\alpha)$ is an isomorphism of rings.

Exercise 3.46. Make a multiplication table for $\mathbb{Z}[\sqrt{3}]/2\mathbb{Z}[\sqrt{3}]$ and determine whether $\mathbb{Z}[\sqrt{3}]/2\mathbb{Z}[\sqrt{3}]$ is a field.

Exercise 3.47. Let $\mathbb{F}$ be a field and let f be a non-constant polynomial in $\mathbb{F}[X]$. Show that there is a field $\mathbb{E}$ containing $\mathbb{F}$, such that $\mathbb{E}$ contains a root of f.

Exercise 3.48. Let R be a ring and let p be a prime number. Show that the function $F : R/pR \to R/pR$ defined by $F(x) \equiv x^p \bmod pR$ is a ring homomorphism. (The function F is called the *Frobenius endomorphism* of R/pR).

Exercise 3.49. Let R be a unique factorization domain (UFD) and let $m \in R$ be irreducible. Show that R/mR is an integral domain. In the case that R is a Euclidean domain, show that R/mR is a field.

3.7 PROOF OF THE SECOND NEBENSATZ

There are many proofs of the Quadratic Reciprocity Law and the Nebensatzen. Many of these proofs are in [13] or [14], and the appendix to [14] contains a list of nearly 200 published proofs. For this book, I've chosen the proof using Gauss sums. There are two reasons for this choice. First, I believe this proof gives the clearest insight into why the theorems are true, and secondly the proofs introduce new and useful concepts. The proof of the Second Nebensatz given in this section will form a blueprint for the proof of the Quadratic Reciprocity Law, which we shall give in the next section.

The ring $\mathbb{Z}[\zeta_8]$. Consider the complex number $\zeta_8 = e^{\frac{2\pi i}{8}}$. By de Moivre's Theorem, we have $\zeta_8 = \frac{1+i}{\sqrt{2}}$, ${\zeta_8}^4 = -1$ and ${\zeta_8}^8 = 1$. The powers of ζ_8 are evenly spaced around the unit circle in $\mathbb{C}$ as follows:

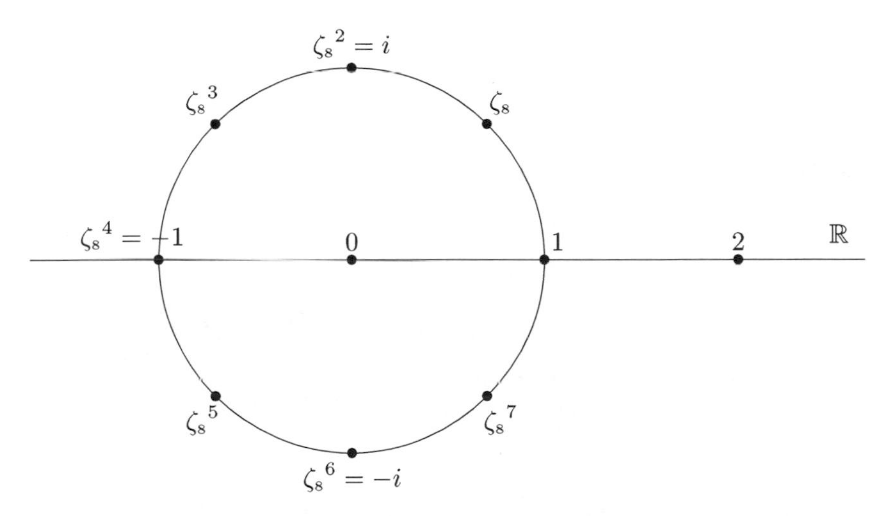

Recall from Section 3.3 that ζ_8 is a primitive eighth root of unity, and therefore ζ_8 is a root of the cyclotomic polynomial Φ_8. In fact, since $\Phi_8(X) = X^4 + 1$, we can see directly that $\Phi_8(\zeta_8) = 0$. In particular, ζ_8 is an algebraic integer. To prove the Second Nebensatz, we shall calculate in the ring $\mathbb{Z}[\zeta_8]$ introduced in Section 3.6. Rings of the form $\mathbb{Z}[e^{\frac{2\pi i}{n}}]$ are called *cyclotomic rings*, so $\mathbb{Z}[\zeta_8]$ is an example of a cyclotomic ring.

The polynomial Φ_8 is irreducible over $\mathbb{Z}$. To see why this is true, we expand out $\Phi_8(X+1)$ using the binomial theorem.

$$\Phi_8(X+1) = X^4 + 4X^3 + 6X^2 + 4X + 2.$$

We see that $\Phi_8(X+1)$ is irreducible, since it satisfies Eisenstein's Criterion with the prime number 2. On the other hand, if $\Phi_8(X+1)$ does not factorize, then neither does $\Phi_8(X)$, and so $\Phi_8(X)$ must also be irreducible.

Since Φ_8 has degree 4, it follows from Corollary 3.6 that elements of $\mathbb{Z}[\zeta_8]$ may be written uniquely in the form

$$a_0 + a_1\zeta_8 + a_2{\zeta_8}^2 + a_3{\zeta_8}^3, \quad \text{with } a_0, a_1, a_2, a_3 \in \mathbb{Z}.$$

Proof of the Second Nebensatz. Let p be an odd prime. Consider the element $G = \zeta_8 - {\zeta_8}^3 = \zeta_8 + {\zeta_8}^{-1}$ in the ring $\mathbb{Z}[\zeta_8]$. We'll calculate the congruency class of G^p modulo $p\mathbb{Z}[\zeta_8]$ in two ways.

(*First calculation*) By Lemma 3.5, we have in any ring R the congruence $(x+y)^p \equiv x^p + y^p \bmod pR$. In particular, this implies

$$G^p \equiv {\zeta_8}^p + {\zeta_8}^{-p} \mod p\mathbb{Z}[\zeta_8].$$

Since ${\zeta_8}^8 = 1$, the complex number ${\zeta_8}^p + {\zeta_8}^{-p}$ depends only on the congruency class of p modulo 8. If p is congruent to ± 1 modulo 8, then we have

$${\zeta_8}^p + {\zeta_8}^{-p} = \zeta_8 + {\zeta_8}^{-1} = G.$$

If on the other hand $p \equiv \pm 3 \bmod 8$, then we have

$${\zeta_8}^p + {\zeta_8}^{-p} = {\zeta_8}^3 + {\zeta_8}^{-3} = -G.$$

This can be summarized as

$$G^p \equiv (-1)^{\frac{p^2-1}{8}} G \mod p\mathbb{Z}[\zeta_8]. \tag{3.3}$$

(*Second calculation*) Note that $G^2 = {\zeta_8}^2 + 2 + {\zeta_8}^{-2} = 2$. Therefore,

$$G^p = G^{p-1}G = 2^{\frac{p-1}{2}}G.$$

By Euler's Criterion, we know that $2^{\frac{p-1}{2}}$ is congruent to $\left(\frac{2}{p}\right)$ modulo $p\mathbb{Z}$ (and therefore also modulo $p\mathbb{Z}[\zeta_8]$). Substituting this into the previous line, we get

$$G^p \equiv \left(\frac{2}{p}\right) G \mod p\mathbb{Z}[\zeta_8]. \tag{3.4}$$

Comparing Equations (3.3) and (3.4), we get

$$\left(\frac{2}{p}\right) G \equiv (-1)^{\frac{p^2-1}{8}} G \mod p\mathbb{Z}[\zeta_8].$$

We would like to cancel G from each side of this congruence. In order to do this, we need to know that G is invertible modulo $p\mathbb{Z}[\zeta_8]$. To see that this is indeed the case, we note that $\frac{p+1}{2}$ is an integer and

$$G \times \left(\frac{p+1}{2}G\right) = p+1 \equiv 1 \mod p\mathbb{Z}[\zeta_8].$$

Therefore, G is invertible modulo $p\mathbb{Z}[\zeta_8]$ with inverse $\frac{p+1}{2}G$. This allows us to cancel G from either side of our congruence, so we get

$$\left(\frac{2}{p}\right) \equiv (-1)^{\frac{p^2-1}{8}} \mod p\mathbb{Z}[\zeta_8].$$

Both sides of the congruence are now either 1 or -1. However, we saw in Equation (3.2) that $1 \not\equiv -1 \bmod p\mathbb{Z}[\zeta_8]$. Therefore, $\left(\frac{2}{p}\right) = (-1)^{\frac{p^2-1}{8}}$. □

3.8 GAUSS SUMS AND THE PROOF OF QUADRATIC RECIPROCITY

The proof of the Second Nebensatz given in Section 3.7 relies on the fact that the number $G = \sqrt{2}$ is an element of the ring $\mathbb{Z}[\zeta_8]$. On the one hand, G^p is related to the quadratic residue symbol $\left(\frac{2}{p}\right)$ by Euler's Criterion. On the other hand, the congruency class of G^p modulo $p\mathbb{Z}[\zeta_8]$ depends only on ${\zeta_8}^p$, which in turn depends only on the congruency class of p modulo 8.

In this section, we shall use a very similar method to prove the Quadratic Reciprocity Law. In particular, we shall show that there is a number $G(p) = \sqrt{\pm p}$, which may be expressed in terms of the primitive p-th root of unity $\zeta_p = e^{\frac{2\pi i}{p}}$. By Euler's Criterion, $G(p)^q$ is related to the quadratic residue symbol $\left(\frac{p}{q}\right)$. On the other hand, the congruency class of $G(p)^q$ modulo $q\mathbb{Z}[\zeta_p]$ depends only on ${\zeta_p}^q$, which in turn depends only on the congruency class of q modulo p.

The cyclotomic rings $\mathbb{Z}[\zeta_p]$. Let p be an odd prime number and consider the complex number

$$\zeta_p = e^{\frac{2\pi i}{p}}.$$

The number ζ_p is a primitive p-th root of unity and is therefore a root of the cyclotomic polynomial Φ_p. In particular, ζ_p is an algebraic integer. The proof of the Quadratic Reciprocity Law is a calculation in the cyclotomic ring $\mathbb{Z}[\zeta_p]$.

Recall from Proposition 3.3 that

$$\Phi_p(X) = \frac{X^p - 1}{X - 1} = 1 + X + \cdots + X^{p-1}.$$

As with the polynomial Φ_8, the polynomial Φ_p is irreducible over $\mathbb{Z}$. To see this, note that

$$\Phi_p(X+1) = \frac{(X+1)^p - 1}{X}.$$

Expanding this out using the binomial theorem, we have

$$\Phi_p(X+1) = \sum_{r=1}^{p} \frac{p!}{r!(p-r)!} X^{r-1}.$$

The leading term of $\Phi_p(X+1)$ is X^{p-1}. By Exercise 3.8, the other coefficients are all multiples of p, and the constant term is p, which is not a multiple of p^2. Therefore, $\Phi_p(X+1)$ satisfies Eisenstein's Criterion with the prime p. In particular, Φ_p is irreducible, and so by Corollary 3.6, the elements of $\mathbb{Z}[\zeta_p]$ can be written uniquely in the form

$$a_0 + a_1\zeta_p + \cdots + a_{p-2}\zeta_p{}^{p-2}$$

with coefficients $a_0, \ldots, a_{p-2}$ in $\mathbb{Z}$.

The Gauss Sum. Let p be an odd prime number. The *Gauss sum* $G(p)$ is defined to be the following element in the cyclotomic ring $\mathbb{Z}[\zeta_p]$:

$$G(p) = \sum_{a=1}^{p-1} \left(\frac{a}{p}\right) \zeta_p{}^a, \quad \text{where } \zeta_p = e^{\frac{2\pi i}{p}}.$$

In this formula, $\left(\frac{a}{p}\right)$ is the quadratic residue symbol. Note that the summand $\left(\frac{a}{p}\right) \zeta_p{}^a$ depends only on the congruency class of a modulo p, and so we can regard the Gauss sum as a sum over $a \in \mathbb{F}_p^{\times}$.

As an example, we'll calculate $G(3)$. By definition, we have

$$G(3) = \left(\frac{1}{3}\right) \zeta_3 + \left(\frac{2}{3}\right) \zeta_3{}^2 = \zeta_3 - \zeta_3{}^2.$$

By De Moivre's theorem, we have $\zeta_3 = -\frac{1}{2} + i\frac{\sqrt{3}}{2}$. This implies $G(3) = i\sqrt{3}$, as is shown in the following picture:

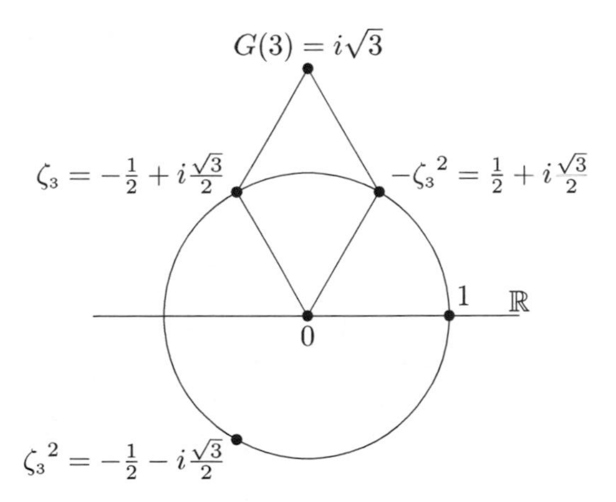

The next lemma shows that $G(p)$ is always a square root of either p or $-p$. This fact will allow are to use $G(p)$ in the same way as we used the number $G = \zeta_8 - \zeta_8{}^3 = \sqrt{2}$ in the proof of the Second Nebensatz.

Lemma 3.7. *For any odd prime p, we have $G(p)^2 = (-1)^{\frac{p-1}{2}} p$.*

Proof. We have

$$G(p)^2 = \left(\sum_{a \in \mathbb{F}_p^\times} \left(\frac{a}{p} \right) \zeta_p{}^a \right) \left(\sum_{b \in \mathbb{F}_p^\times} \left(\frac{b}{p} \right) \zeta_p{}^b \right).$$

Expanding out the brackets, we get

$$G(p)^2 = \sum_{a \in \mathbb{F}_p^\times} \left(\sum_{b \in \mathbb{F}_p^\times} \left(\frac{ab}{p} \right) \zeta_p{}^{a+b} \right).$$

We shall change the variable in the inner sum from b to c, where $b \equiv ac \bmod p$. Note that since a is invertible modulo p, the variable c runs through $\mathbb{F}_p^\times$ as b runs through $\mathbb{F}_p^\times$. With this new notation, we have

$$G(p)^2 = \sum_{a \in \mathbb{F}_p^\times} \left(\sum_{c \in \mathbb{F}_p^\times} \left(\frac{a^2 c}{p} \right) \zeta_p{}^{a+ac} \right).$$

Swapping the order of summation and using the fact that $\left(\frac{a^2}{p}\right) = 1$, we get

$$G(p)^2 = \sum_{c \in \mathbb{F}_p^\times} \left(\frac{c}{p} \right) \left(\sum_{a=1}^{p-1} \zeta_p{}^{(1+c)a} \right). \tag{3.5}$$

We shall evaluate the inner sum over the variable a in Equation (3.5). This inner sum is a geometric progression whose terms have ratio $\zeta_p{}^{1+c}$.

If $c \not\equiv -1 \bmod p$, then ${\zeta_p}^{1+c} \neq 1$, so we have (using the formula for a geometric progression)

$$\sum_{a=1}^{p-1} {\zeta_p}^{(1+c)a} = \frac{{\zeta_p}^{(1+c)p} - {\zeta_p}^{1+c}}{{\zeta_p}^{1+c} - 1} = -1 \quad (c \not\equiv -1 \bmod p).$$

On the other hand, when $c \equiv -1 \bmod p$, we have ${\zeta_p}^{1+c} = 1$. In this case, every term in the inner sum is 1, so we have

$$\sum_{a=1}^{p-1} {\zeta_p}^{(1+c)a} = \sum_{a=1}^{p-1} 1 = p - 1 \quad (c \equiv -1 \bmod p).$$

Substituting these values into Equation (3.5), we get

$$G(p)^2 = -\sum_{c=1}^{p-1} \left(\frac{c}{p}\right) + p\left(\frac{-1}{p}\right). \tag{3.6}$$

Recall that by Corollary 3.3, there are the same number of quadratic residues as quadratic non-residues modulo p. This implies

$$\sum_{c=1}^{p-1} \left(\frac{c}{p}\right) = 0.$$

Substituting this into Equation (3.6) and using the First Nebensatz, we're left with

$$G(p)^2 = (-1)^{\frac{p-1}{2}} p.$$

□

Theorem 3.10 (Quadratic Reciprocity Law). *For any two distinct odd primes p and q, we have*

$$\left(\frac{p}{q}\right) = (-1)^{\frac{(p-1)(q-1)}{4}} \left(\frac{q}{p}\right).$$

Proof. Let p and q be distinct odd primes. As with the proof of the Second Nebensatz, we'll calculate $G(p)^q$ modulo $q\mathbb{Z}[\zeta_p]$ by two different methods and then compare the results.

(*First calculation*) We've seen in Lemma 3.5 that the following congruence holds in any ring R:

$$(x + y)^q \equiv x^q + y^q \mod qR.$$

Applying this congruence to $G(p)^q$ modulo $q\mathbb{Z}[\zeta_p]$, we have

$$G(p)^q \equiv \sum_{a \in \mathbb{F}_p^\times} \left(\left(\frac{a}{p} \right) {\zeta_p}^a \right)^q \mod q\mathbb{Z}[\zeta_p].$$

Since q is odd, we have $\left(\frac{a}{p}\right)^q = \left(\frac{a}{p}\right)$, so we can rewrite this as

$$G(p)^q \equiv \sum_{a \in \mathbb{F}_p^\times} \left(\frac{a}{p} \right) {\zeta_p}^{aq} \mod q\mathbb{Z}[\zeta_p].$$

We'll change the variable in this sum from a to b, where $b \equiv qa \bmod p$. Note that as q is invertible modulo p, the variable b runs through $\mathbb{F}_p^\times$ as a does.

$$G(p)^q \equiv \sum_{b \in \mathbb{F}_p^\times} \left(\frac{bq^{-1}}{p} \right) {\zeta_p}^b \mod q\mathbb{Z}[\zeta_p].$$

Since $\left(\frac{q^{-1}}{p}\right) = \left(\frac{q}{p}\right)$, we have

$$G(p)^q \equiv \left(\frac{q}{p} \right) \sum_{b \in \mathbb{F}_p^\times} \left(\frac{b}{p} \right) {\zeta_p}^b \equiv \left(\frac{q}{p} \right) G(p) \mod q\mathbb{Z}[\zeta_p]. \tag{3.7}$$

(*Second calculation*) Since q is odd, we can write $G(p)^q$ in the form

$$G(p)^q = G(p) \cdot \left(G(p)^2 \right)^{\frac{q-1}{2}}.$$

We saw in Lemma 3.7 that $G(p)^2 = (-1)^{\frac{p-1}{2}} p$. Substituting this into $G(p)^q$, we get

$$G(p)^q = G(p) \cdot (-1)^{\frac{(p-1)(q-1)}{4}} \cdot p^{\frac{q-1}{2}}.$$

By Euler's Criterion, the term $p^{\frac{q-1}{2}}$ is congruent modulo q (and hence modulo $q\mathbb{Z}[\zeta_p]$) to the quadratic residue symbol $\left(\frac{p}{q}\right)$, so we have

$$G(p)^q \equiv G(p) \cdot (-1)^{\frac{(p-1)(q-1)}{4}} \left(\frac{p}{q} \right) \mod q\mathbb{Z}[\zeta_p]. \tag{3.8}$$

Comparing Equations (3.7) and (3.8), we get

$$G(p) \cdot \left(\frac{q}{p} \right) \equiv G(p) \cdot (-1)^{\frac{(p-1)(q-1)}{4}} \left(\frac{p}{q} \right) \mod q\mathbb{Z}[\zeta_p].$$

In order to cancel $G(p)$ from both sides, we need to show that $G(p)$ is invertible modulo $q\mathbb{Z}[\zeta_p]$. For this, let p^{-1} be the inverse of p modulo q.

Since $G(p)^2 = \pm p$, we have

$$G(p) \cdot \left(G(p)p^{-1}\right) \equiv \pm 1 \mod q\mathbb{Z}[\zeta_p].$$

We can therefore cancel $G(p)$ to get

$$\left(\frac{p}{q}\right) \equiv (-1)^{\frac{(p-1)(q-1)}{4}} \left(\frac{q}{p}\right) \mod q\mathbb{Z}[\zeta_p].$$

Since both sides are either 1 or -1, we have by Equation (3.2)

$$\left(\frac{p}{q}\right) = (-1)^{\frac{(p-1)(q-1)}{4}} \left(\frac{q}{p}\right).$$

□

Exercise 3.50. Show that for any odd prime p, the Gauss sum is given by

$$G(p) = \sum_{x=0}^{p-1} \zeta_p^{x^2}.$$

Exercise 3.51. Let p and q be odd primes such that $p \equiv 1 \bmod 4$ and $q \equiv 1 \bmod p$, and let g be a primitive root modulo q. Show that the integer

$$x = \sum_{a=1}^{p-1} \left(\frac{a}{p}\right) g^{\frac{(q-1)a}{p}}$$

is a solution to the congruence $x^2 \equiv p \bmod q$.

Exercise 3.52. Let $f(x)$ be a polynomial with coefficients in $\mathbb{Z}$. Show that if f has a root in $\mathbb{Z}[\zeta_p]$, then for every prime $q \equiv 1 \bmod p$, the congruence $f(x) \equiv 0 \bmod q$ has a solution.

Exercise 3.53. Let $f(x) = x^3 + x^2 - 2x - 1$. Show that for every prime $p \equiv 1 \bmod 7$, the congruence $f(x) \equiv 0 \bmod p$ has a solution. Verify that a solution exists modulo 29.

Exercise 3.54. Calculate the Gauss sums $G(p)$ for several primes p. Hence, make a conjecture about the value of $G(p)$. Recall that $G(p)^2 = (-1)^{\frac{p-1}{2}} p$, so this is just a matter of deciding which square root we take. For example, one can define a function `G(p)` in `sage`, which calculates the Gauss sum as a complex number (as opposed to an element of the cyclotomic ring):

```
def G(p):
    zeta=exp(2*pi*i/p)
    return CC(sum(zeta^(x^2), x ,0,p-1))
```

```
G(11)
    2.23606797749976 + 6.43929354282591e-15*I
```

(Proving the conjecture is difficult, but you can look up the proof in Section 6.4 of the book, *A Classical Introduction to Modern Number Theory* [13].)

Exercise 3.55. Let $p \geq 3$ be a prime number and $\zeta_p = e^{\frac{2\pi i}{p}}$. Show that for $a = 1, \ldots, p-2$ the number $u = 1 + \zeta_p + \cdots + \zeta_p{}^a$ is invertible in the ring $\mathbb{Z}[\zeta_p]$, i.e. show that $\frac{1}{u} \in \mathbb{Z}[\zeta_p]$.

HINTS FOR SOME EXERCISES

3.6. Use the Chinese Remainder Theorem.

3.7. Show that $F(x) = \sum_{a \in \mathbb{F}_p} F(a)\left(1 - (x-a)^{p-1}\right)$. Then count how many functions there are, and how many polynomials of degree $\leq p-1$.

3.8. In the last part, use induction on x.

3.13. In the last part of the question, use the fact that the series $\sum \frac{1}{n}$ diverges.

3.18. Use the Chinese Remainder Theorem.

3.20. Show that if a is a primitive root, then a^{-1} is a primitive root.

3.21. Use Proposition 3.3.

3.33. Show that $\left(\frac{a}{p}\right) = 1$ for $1 \leq a \leq N$.

3.36. In the first part of the question, you could use the First Nebensatz. In the second part, let $N = 4(p_1 \cdots p_r)^2 + 1$.

3.40. In the second part of the question, let $\chi(p) = \left(\frac{n}{p}\right)$ for odd primes p which do not divide n. Use the Reciprocity Law, the Chinese Remainder Theorem and also the first part of the question.

3.42. Look back at the proof of the Chinese Remainder Theorem.

3.47. Let $\mathbb{E} = \mathbb{F}[X]/p\mathbb{F}[X]$, where p is an irreducible factor of f.

3.51. Interpret Lemma 3.7 as a congruence in $\mathbb{Z}[X]/\Phi_p\mathbb{Z}[X]$ rather than an equation in $\mathbb{Z}[\zeta_p]$. Alternatively, modify the proof of the Lemma 3.7.

3.52. Suppose the root in $\mathbb{Z}[\zeta_p]$ is $\sum a_i {\zeta_p}^i$. Take $x = \sum a_i g^{\frac{(q-1)i}{p}}$, where g is a primitive root modulo q.

3.53. Show that $f(2\cos(\frac{2\pi}{7})) = 0$ and then use Exercise 3.52.

Chapter 4

p-Adic Methods in Number Theory

Roughly speaking, p-adic methods are methods for dealing with congruences modulo p^n for large n, where p is a prime. In Chapter 3, we studied congruences modulo a prime number, and we made great use of the uniqueness of factorizations in the polynomial ring $\mathbb{F}_p[X]$. The methods needed for studying congruences modulo p^n are very different because polynomials over $\mathbb{Z}/p^n$ do not factorize uniquely. For example over $\mathbb{Z}/8$, the polynomial $X^2 - 1$ has two different factorizations:

$$X^2 - 1 \equiv (X+1)(X-1) \equiv (X+3)(X-3) \bmod 8.$$

From this, we see that $X^2 - 1$ has four roots in $\mathbb{Z}/8$, even though it only has degree 2.

Instead, p-adic methods tend to be based on ideas from calculus. In Sections 4.1 and 4.2, we'll solve congruences by a method based on the Newton–Raphson algorithm. In Sections 4.3 and 4.4, we'll use power series methods. There are places (the proof of Lemma 4.2 and also Section 4.6) where some knowledge of real analysis would help when reading this chapter.

4.1 HENSEL'S LEMMA

The Newton–Raphson Method. For a moment we'll discuss the Newton–Raphson method from calculus, which at first sight has nothing at all to do with number theory.

The Newton–Raphson method is a way of finding an approximate solution to an equation $f(x) = 0$, where $f : \mathbb{R} \to \mathbb{R}$ is a differentiable

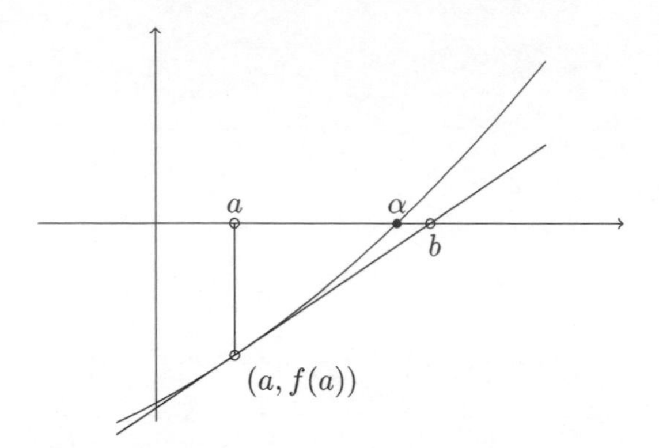

Figure 4.1: First step of the Newton–Raphson method.

function. Assume that there is a solution $x = \alpha$ to the equation $f(x) = 0$, and we'd like to find a good estimate for α. For example, we may be trying to estimate π by solving the equation $\sin(x) = 0$.

We begin with a real number a, which is fairly close to α. In order to find a number b even closer to α, we draw the tangent line to the graph $y = f(x)$ at the point $(a, f(a))$. This tangent line is quite close to the graph, and so it should cross the x-axis near where the graph crosses the x-axis, i.e. close to the solution α (see Figure 4.1). The tangent line has the following equation:

$$y = f(a) + f'(a) \cdot (x - a).$$

The point $(b, 0)$, where this line crosses the x-axis, is determined by setting $y = 0$:

$$0 = f(a) + f'(a)(b - a).$$

Solving this equation we can express b in terms of a:

$$b = a - \frac{f(a)}{f'(a)}.$$

To summarize the discussion above, if a is fairly close to the solution α, then we expect that b is even closer. We may then repeat the process from the beginning with a replaced by b, to find an even better approximation $c = b - \frac{f(b)}{f'(b)}$ etc. In this way, we obtain a sequence of real numbers $a_0 = a$, $a_1 = b$, where each a_{n+1} is defined recursively in terms of a_n by the formula

$$a_{n+1} = a_n - \frac{f(a_n)}{f'(a_n)}. \tag{4.1}$$

We expect that the sequence a_n converges to the root α (see Figure 4.2).

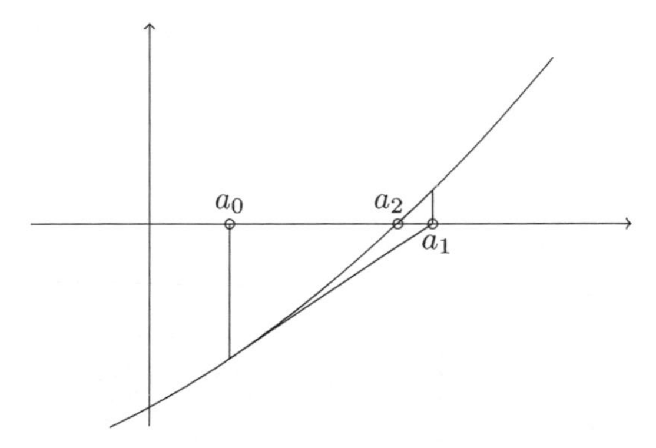

Figure 4.2: Second step of the Newton–Raphson method.

Example. We'll find an approximate value of $\sqrt{2}$ using the Newton–Raphson method. To do this, we regard $\sqrt{2}$ as a root of the polynomial $f(X) = X^2 - 2$. We'll begin with the number $a_0 = 1$, which is fairly close to $\sqrt{2}$, and then calculate the next few terms of the sequence a_n. Note that we have

$$a_{n+1} = a_n - \frac{a_n^2 - 2}{2a_n} = \frac{a_n}{2} + \frac{1}{a_n}.$$

The next few terms are:

$$a_1 = \frac{1}{2} + \frac{1}{1} = \frac{3}{2} = 1.5,$$

$$a_2 = \frac{3}{4} + \frac{2}{3} = \frac{17}{12} \approx 1.416667,$$

$$a_3 = \frac{17}{24} + \frac{12}{17} = \frac{577}{488} \approx 1.41421568,$$

$$a_4 = \frac{577}{976} + \frac{488}{577} = \frac{665857}{470832} \approx 1.41421356237469.$$

These estimates are fairly good, as the value of $\sqrt{2}$ (up to 14 decimal places) is 1.41421356237310. The sequence a_n converges very rapidly to $\sqrt{2}$; the number of accurate significant figures roughly doubles with each iteration.

This process can be automated on `sage` as follows

```
var('X')
R.<X>=RR[X]
f=X^2-2
f.newton_raphson(5,1)
```

[1.50000000000000, 1.41666666666667, 1.41421568627451,
1.41421356237469, 1.41421356237310]

The first two lines of code introduce a new symbol `X` and define `R` to be the ring of polynomials over $\mathbb{R}$ in the variable `X`. If we'd like the approximations given as rational numbers, then we must change the second line to `R.<X>=QQ[X]`, which sets $R = \mathbb{Q}[X]$ instead of $\mathbb{R}[X]$.

Solving congruences by the Newton–Raphson method. A method similar to the Newton–Raphson method allows us to solve congruences. Suppose we have a polynomial f with integer coefficients, and we'd like to find a solution to the congruence $f(x) \equiv 0 \bmod p^N$, where p^N is a large power of a prime number p. We can try the following:

(1) Find an integer a_0 such that $f(a_0) \equiv 0 \bmod p^r$, where r is a small number (we can think of a_0 as an "approximate root").
(2) Recursively define a sequence of rational numbers a_n by the Newton–Raphson formula

$$a_{n+1} = a_n - \frac{f(a_n)}{f'(a_n)}.$$

It often happens that for n sufficiently large, $f(a_n) \equiv 0 \bmod p^N$. The main result of this section is Hensel's Lemma, which is a criterion for this method to work. Before describing the theory, we'll give an example.

Example. Let $f(X) = X^2 + 2$ and let $p = 3$. This means we are trying to find a square root of -2 modulo 3^N. We begin by choosing an approximate root $a_0 = 1$. Note that $f(a_0) = 1^2 + 2 \equiv 0 \bmod 3$, so a_0 is a root of f modulo 3. The recursive formula for the sequence a_n is

$$a_{n+1} = a_n - \frac{a_n^2 + 2}{2a_n} = \frac{a_n}{2} - \frac{1}{a_n}.$$

The next two terms in the sequence are:

$$\begin{aligned} a_1 &= \frac{1}{2} - 1 = -\frac{1}{2}, \\ a_2 &= -\frac{1}{4} + 2 = \frac{7}{4}. \end{aligned}$$

We'll see that a_1 is a solution to $f(x) \equiv 0 \bmod 3^2$ and a_2 is a solution to $f(x) \equiv 0 \bmod 3^4$. This might seem a strange thing to say, since neither a_1 nor a_2 is an integer. To interpret a rational number $\frac{n}{m}$ as a congruency class, we can replace it by $n \times m^{-1}$, as long as m is invertible. For example, to interpret $a_1 = -\frac{1}{2}$ as a congruency class modulo 9, we calculate $2^{-1} \equiv -4 \bmod 9$, and therefore

$$a_1 \equiv 4 \mod 9.$$

Similarly, since $4^{-1} \equiv -20 \bmod 81$, we have

$$a_2 \equiv 7 \times -20 \equiv -140 \equiv 22 \mod 81.$$

We can easily check that $4^2 \equiv -2 \bmod 9$ and $22^2 \equiv -2 \bmod 81$, so a_1 and a_2 are indeed solutions to the congruence modulo 9 and modulo 81, respectively.

The Local Ring. In the example above, we have reduced rational numbers modulo p^n, rather than just integers. A rational number $\frac{a}{b}$ can be reduced modulo p^n as long as its denominator b is invertible modulo p^n. Recall that b is invertible modulo p^n if and only if it is coprime to p^n, and this is the case if and only if b is not a multiple of p. We can therefore make sense of $\frac{a}{b}$ modulo p^n as long as b is not a multiple of p. We'll write $\mathbb{Z}_{(p)}$ for the set of rational numbers with this property, i.e.

$$\mathbb{Z}_{(p)} = \left\{ \frac{a}{b} : a, b \in \mathbb{Z} \text{ and } b \text{ is not a multiple of } p \right\}.$$

For example, $\frac{2}{3}$ is in $\mathbb{Z}_{(5)}$ but $\frac{3}{10}$ is not in $\mathbb{Z}_{(5)}$ since its denominator is a multiple of 5. We call $\mathbb{Z}_{(p)}$ the *local ring*[a] at p.

Proposition 4.1. *$\mathbb{Z}_{(p)}$ is a subring of $\mathbb{Q}$.*

Proof. Clearly 0 and 1 are in $\mathbb{Z}_{(p)}$, so by Exercise 1.16 we need only show that $\mathbb{Z}_{(p)}$ is closed under addition, subtraction and multiplication. Let $x = \frac{a}{b}$ and $y = \frac{c}{d}$ be elements of $\mathbb{Z}_{(p)}$, so neither b nor d is a multiple of p. Then, we have

$$x + y = \frac{ad + bc}{bd}, \qquad x - y = \frac{ad - bc}{bd}, \qquad xy = \frac{ac}{bd}.$$

By Euclid's Lemma, the denominator bd is not a multiple of p. Therefore $x + y$, $x - y$ and xy are all in $\mathbb{Z}_{(p)}$. □

We now seem to have two different notions of congruence. For integers x and y, we say that x and y are congruent if $x - y$ is an integer multiple of p^n.

---

[a]In general, a "local ring" is a ring with a unique maximal ideal, or in more elementary terms a ring in which the non-invertible elements form an additive subgroup. The non-invertible elements of $\mathbb{Z}_{(p)}$ are the multiples of p.

However, for elements $x, y \in \mathbb{Z}_{(p)}$, recall that $x \equiv y \bmod p^n\mathbb{Z}_{(p)}$ means that $x - y = p^n z$ for some $z \in \mathbb{Z}_{(p)}$. The following lemma shows that in fact these two notions are the same.

Lemma 4.1. *For elements $\frac{a}{b}$ and $\frac{c}{d}$ in $\mathbb{Z}_{(p)}$, we have $\frac{a}{b} \equiv \frac{c}{d} \bmod p^n\mathbb{Z}_{(p)}$ if and only if $a \cdot b^{-1} \equiv c \cdot d^{-1} \bmod p^n$.*

Proof. Suppose $\frac{a}{b} \equiv \frac{c}{d} \bmod p^n\mathbb{Z}_{(p)}$. This means that there is an element $\frac{e}{f} \in \mathbb{Z}_{(p)}$ such that $\frac{a}{b} - \frac{c}{d} = p^n \frac{e}{f}$. Multiplying by all the denominators, we get $adf - cbf = p^n ebd$, and in particular $adf \equiv cbf \bmod p^n$. Since b, d and f are invertible modulo p^n, we have $ab^{-1} \equiv cd^{-1} \bmod p^n$.

Conversely, suppose $ab^{-1} \equiv cd^{-1} \bmod p^n$. This implies $ad \equiv bc \bmod p^n$, so there is an integer e such that $ad - bc = p^n e$. Hence $\frac{a}{b} - \frac{c}{d} = p^n \frac{e}{bd}$, and so $\frac{a}{b} \equiv \frac{c}{d} \bmod p^n\mathbb{Z}_{(p)}$. □

The lemma shows that congruences modulo $p^n\mathbb{Z}_{(p)}$ are equivalent to congruences modulo p^n, and so in this chapter we shall often write $x \equiv y \bmod p^n$, when we actually mean $x \equiv y \bmod p^n\mathbb{Z}_{(p)}$; the numbers x and y need not be integers, but they represent congruency classes of integers.

Hensel's Lemma. The main theorem of this section is Hensel's Lemma, which is a criterion for the Newton–Raphson method to succeed in solving congruences. This lemma is most easy to state if we introduce some more notation. For a prime p, let $v_p(n)$ be the largest integer a such that p^a is a factor of n. We define $v_p(0) = \infty$; this makes sense since 0 is a multiple of p^a for every a. The number $v_p(n)$ is called *the p-adic valuation of n*. For example, we have

$$v_2(24) = 3, \quad v_3(24) = 1, \quad v_5(24) = 0.$$

Since factorization in $\mathbb{Z}$ is unique, we have

$$v_p(nm) = v_p(n) + v_p(m).$$

More generally, if $\frac{n}{m}$ is a rational number then we define $v_p(\frac{n}{m}) = v_p(n) - v_p(m)$. With this notation, there is an alternative description of the local ring:

$$\mathbb{Z}_{(p)} = \{x \in \mathbb{Q} : v_p(x) \geq 0\}. \tag{4.2}$$

More generally, for a positive integer r, the inequality $v_p(x) \geq r$ is equivalent to the congruence $x \equiv 0 \bmod p^r$.

Theorem 4.1 (Hensel's Lemma). *Let p be a prime number and let f be a polynomial with coefficients in $\mathbb{Z}_{(p)}$. Suppose there is an $a_0 \in \mathbb{Z}_{(p)}$ such that*

$$f(a_0) \equiv 0 \bmod p^{2c+1}, \quad \textit{where } c = v_p(f'(a_0)).$$

Define a sequence (a_n) recursively by $a_{n+1} = a_n - \frac{f(a_n)}{f'(a_n)}$. Then the numbers a_n are all in $\mathbb{Z}_{(p)}$ and

$$f(a_n) \equiv 0 \mod p^{2c+2^n}.$$

Example. Let $p = 2$ and $f(X) = X^2 + 7$. We'll check that $a_0 = 1$ satisfies the conditions of Hensel's Lemma. Note that $f'(X) = 2X$ and therefore $f'(a_0) = 2$. The number c in Hensel's Lemma is given by

$$c = v_2(2) = 1.$$

Since $2^{2c+1} = 8$, we need only check that a_0 is a root of f modulo 8. This is indeed true, because $f(1) = 8 \equiv 0 \bmod 8$. We can therefore find roots of f modulo higher powers of 2. The recursive formula for these roots is:

$$a_{n+1} = a_n - \frac{a_n^2 + 7}{2a_n} = \frac{a_n}{2} - \frac{7}{2a_n}.$$

The next few terms in the sequence are:

$$\begin{aligned} a_1 &= \frac{1}{2} - \frac{7}{2} &&= -\frac{6}{2} &&= -3, \\ a_2 &= -\frac{3}{2} + \frac{7}{6} &&= -\frac{2}{6} &&= -\frac{1}{3}, \\ a_3 &= -\frac{1}{6} + \frac{21}{2} &&= \frac{62}{6} &&= \frac{31}{3}. \end{aligned}$$

We can check that these are solutions to the congruence

$$\begin{aligned} a_1^2 + 7 &= 9 + 7 = 16 &&\equiv 0 \mod 2^4, \\ a_2^2 + 7 &= \frac{1}{9} + 7 = \frac{64}{7} &&\equiv 0 \mod 2^6, \\ a_3^2 + 7 &= \frac{961}{9} + 7 = \frac{1024}{9} &&\equiv 0 \mod 2^{10}. \end{aligned}$$

So a_3 is a square root of -7 modulo 2^{10}. Note that we can write a_3 modulo 2^{10} more explicitly:

$$a_3 \equiv 31 \times 3^{-1} \equiv 693 \mod 1024.$$

Of course, we can check 693 is genuinely an integer solution to our congruence modulo 1024:

$$693^2 + 7 = 480256 \equiv 0 \mod 1024.$$

Proof of Hensel's Lemma. We'll prove Hensel's Lemma by induction on n. In fact, it will be easier to prove that all three of the following statements are true for every n:

(1) $a_n \in \mathbb{Z}_{(p)}$ and $a_n \equiv a_0 \bmod p^{c+1}$;
(2) $v_p(f'(a_n)) = c$;
(3) $f(a_n) \equiv 0 \bmod p^{2c+2^n}$.

Hensel's Lemma requires only parts 1 and 3 to be proved, but we cannot make the inductive step work without also using part 2.

If $n = 0$, then parts 1, 2 and 3 are the conditions on a_0 which we are assuming. Assume that statements 1, 2 and 3 are all true for a_n; we'll prove them for a_{n+1}.

(1) We'll use the formula

$$a_{n+1} = a_n - \delta, \quad \text{where} \quad \delta = \frac{f(a_n)}{f'(a_n)}.$$

By the inductive hypotheses for a_n, we have

$$v_p(\delta) = \underbrace{v_p\big(f(a_n)\big)}_{\geq 2c+2^n \text{ by } (3)} - \underbrace{v_p\big(f'(a_n)\big)}_{=c \text{ by } (2)} \geq c + 2^n. \tag{4.3}$$

In particular, Equation (4.3) shows that $v_p(\delta) \geq 0$, which means $\delta \in \mathbb{Z}_{(p)}$. This implies $a_{n+1} \in \mathbb{Z}_{(p)}$. Furthermore, we see from Equation (4.3) that $v_p(\delta) \geq c + 1$. We can interpret this inequality as a congruence: $a_{n+1} \equiv a_n \mod p^{c+1}$. By the inductive hypothesis, we have $a_n \equiv a_0 \bmod p^{c+1}$, and so it follows that

$$a_{n+1} \equiv a_0 \mod p^{c+1}.$$

This proves (1) for a_{n+1}.

(2) Recall that $v_p(f'(a_0)) = c$. This means that $f'(a_0)$ is a multiple of p^c, but is not a multiple of p^{c+1}. Therefore,

$$f'(a_0) \equiv 0 \bmod p^c, \quad f'(a_0) \not\equiv 0 \bmod p^{c+1}.$$

On the other hand, we've shown in (1) that $a_{n+1} \equiv a_0 \bmod p^{c+1}$. This implies

$$f'(a_{n+1}) \equiv f'(a_0) \mod p^{c+1}.$$

In particular, $f'(a_{n+1}) \equiv 0 \bmod p^c$ and $f'(a_{n+1}) \not\equiv 0 \bmod p^{c+1}$. This shows that $v_p(f'(a_{n+1})) = c$, which proves (2) for a_{n+1}.

(3) From Equation (4.3) we have $v_p(\delta) \geq c+2^n$, and therefore $v_p(\delta^2) \geq 2c + 2^{n+1}$. We can interpret this inequality as a congruence:

$$\delta^2 \equiv 0 \mod p^{2c+2^{n+1}}.$$

Hence, in the binomial expansion of $(a_n - \delta)^r$ modulo $p^{2c+2^{n+1}}$, there are only two non-zero terms; the other terms are multiplies of δ^2.

$$a_{n+1}^r = (a_n - \delta)^r \equiv a_n^r - ra_n^{r-1}\delta \mod p^{2c+2^{n+1}}. \tag{4.4}$$

Suppose the polynomial f is given by

$$f(X) = \sum c_r X^r.$$

Substituting a_{n+1} into f and using Equation (4.4), we get

$$\begin{aligned} f(a_{n+1}) &\equiv \sum c_r \left(a_n^r - ra_n^{r-1}\delta\right) && \mod p^{2c+2^{n+1}} \\ &\equiv \sum c_r a_n^r - \left(\sum c_r r a_n^{r-1}\right)\delta && \mod p^{2c+2^{n+1}} \\ &\equiv f(a_n) - f'(a_n)\delta && \mod p^{2c+2^{n+1}}. \end{aligned}$$

Since $\delta = \frac{f(a_n)}{f'(a_n)}$, it follows that $f(a_{n+1}) \equiv 0 \bmod p^{2c+2^{n+1}}$, which proves (3). □

Speeding up calculations. When solving congruences using Hensel's lemma, we calculated a sequence a_n of elements in the local ring $\mathbb{Z}_{(p)}$. We showed that a_n is a solution to the congruence $f(x) \equiv 0 \bmod p^{2c+2^n}$. However, the rational numbers a_n can quickly get complicated. In particular, it can take a little time to invert the denominator of a_n modulo p^{2c+2^n}. There are certain tricks for doing these calculations more quickly. These tricks roughly double the speed of our calculations.

Suppose that we have calculated a_5, which is a solution to $f(a_5) \equiv 0 \bmod p^{2c+2^5}$. If we choose another number $b_5 \equiv a_5 \bmod p^{2c+2^5}$, then it is clear that b_5 satisfies all three inductive hypotheses in the proof of Hensel's Lemma with $n = 5$. Therefore, if we define a sequence b_n, for $n \geq 5$ by the Newton–Raphson formula, then each b_n will also be a root of f modulo 2^{2c+2^n}. What this means, is that at each stage, rather than calculating the number a_n in the local ring, it is sufficient to calculate its congruency class modulo p^{2c+2^n}. In fact by being a little more careful, one can show that we need only calculate a_n modulo p^{c+2^n}.

As an example, let's return to the polynomial $f(X) = X^2 + 2$, which we considered earlier modulo powers of 3. We began with the number $a_0 = 1$, which is a root of f modulo 3. From this, we calculate

$$a_1 = a_0 - \frac{f(a_0)}{f'(a_0)}.$$

We know that a_1 will be a root of f modulo 9, so we'll calculate the congruency class of a_1 modulo 9. This is given by

$$a_1 \equiv 1 - \frac{3}{2} \mod 9.$$

It appears that we need to calculate the inverse of 2 modulo 9. However, this is not the case; we only need to calculate $2^{-1} \bmod 3$. The reason is because we are about to multiply this inverse by 3. Since $2^{-1} \equiv -1 \bmod 3$, we have $3 \cdot 2^{-1} \equiv -3 \bmod 9$. From this, we see that

$$a_1 \equiv 4 \mod 9.$$

Next we calculate a_2 modulo 81:

$$a_2 \equiv 4 - \frac{18}{8} \mod 81.$$

Again, it appears that we need to calculate $8^{-1} \bmod 81$. However, this is not the case, because we will multiply this answer by 18. Since $8^{-1} \equiv -1 \bmod 9$, it follows that $18 \cdot 8^{-1} \equiv -18 \bmod 81$. This implies

$$a_2 \equiv 22 \mod 81.$$

Similarly we can calculate a_3 modulo 3^8 quite easily:

$$a_3 \equiv 22 - \frac{f(22)}{44} \mod 3^8.$$

It appears that we need to calculate $44^{-1} \bmod 3^8$. However, we'll multiply this inverse by $f(22)$, which we know to be a multiple of 3^4. Therefore, we only need to know the inverse of 44 modulo 3^4. A short calculation shows that $44^{-1} \equiv 35 \bmod 81$, and so we have

$$a_3 \equiv 22 - 486 \cdot 35 \equiv 3866 \mod 3^8.$$

As before, the number 3866 is a root of f modulo 3^8.

Throughout this discussion, we've been using the fact that if $x \equiv y \bmod p^r$, then $p^s x \equiv p^s y \bmod p^{r+s}$. This is easy to prove (see Exercise 4.2).

One might wonder whether replacing a_n by another number $b_n \equiv a_n \bmod p^{c+2^n}$ might lead to a different solution to the congruence modulo higher powers of p. In fact, this is not the case. One can check that if

$a_n \equiv b_n \bmod p^{c+2^n}$ then $a_{n+1} \equiv b_{n+1} \bmod p^{c+2^{n+1}}$. Indeed if a is any root of f modulo p^{2c+2^n}, such that $a \equiv a_0 \bmod p^{c+1}$, then one can show that $a \equiv a_n \bmod p^{c+2^n}$.

Exercise 4.1. Find a solution to each of the following congruences:

$$\begin{aligned} x^2 &\equiv 15 \bmod 7^4, & x^2 &\equiv 5 \mod 11^4, \\ x^2 &\equiv 3 \mod 11^4, & x^2 &\equiv -1 \bmod 5^4, \\ x^2+x+5 &\equiv 0 \mod 7^4, & x^3 &\equiv 3 \mod 5^4. \end{aligned}$$

The answers can be checked on `sage` using the `solve_mod` command. For example,

```
solve_mod(x^2==15,7^4)
```

$[(155,),(2246,)]$

Another method is to use the `newton_raphson` function for polynomials over the ring $\mathbb{Z}/7^4$, starting with the approximate root $a_0 = 1$ and calculating a_1, a_2, a_3, a_4.

```
var('X')
R.<X>=IntegerModRing(7^4)[X]
f=X^2-15
f.newton_raphson(4,1)
```

[8, 155, 155, 155]

Exercise 4.2. Show that if $x \equiv y \bmod p^r$, then $p^s x \equiv p^s y \bmod p^{r+s}$.

Exercise 4.3. Show that the units in $\mathbb{Z}_{(p)}$ are the rational numbers x such that $v_p(x) = 0$. Show that the irreducible elements of $\mathbb{Z}_{(p)}$ are the rational numbers x such that $v_p(x) = 1$.

Exercise 4.4. Show that for any positive integer n,

$$n = \prod_p p^{v_p(n)},$$

where the product is taken over all prime numbers p. Show that for positive integers n and m, we have

$$\operatorname{hcf}(n,m) = \prod_p p^{\min(v_p(n),v_p(m))}, \quad \operatorname{lcm}(n,m) = \prod_p p^{\max(v_p(n),v_p(m))}.$$

Hence, show that $nm = \operatorname{hcf}(n,m) \cdot \operatorname{lcm}(n,m)$.

Exercise 4.5. Let p be a prime number, and assume that $p \neq 7$. Consider the cyclotomic polynomial $\Phi_7(X) = 1+X+X^2+X^3+X^4+X^5+X^6$. Show that $\Phi_7(x) \equiv 0 \bmod p^{1000}$ has solutions if and only if $p \equiv 1 \bmod 7$.

Exercise 4.6. Let $n > 0$. Show that for every $y \in \mathbb{Z}/7^n$ there is a unique $x \in \mathbb{Z}/7^n$ such that $x^7 + 14x^5 + 28x^4 + 7x^3 + 49x^2 + 96x + 62 \equiv y \mod 7^n$.

The next exercise shows that Hensel's Lemma works well for finding simple roots, but is not good at finding repeated roots of polynomials.

Exercise 4.7. Let p be a prime number and let $f \in \mathbb{Z}_{(p)}[X]$. By an *approximate root of* f, we mean an element $a \in \mathbb{Z}_{(p)}$ such that $f(a) \neq 0$, but $f(a) \equiv 0 \bmod p^{2c+1}$, where $c = v_p(f'(a))$.

(1) Let $f(X)$ be a polynomial over $\mathbb{Z}_{(p)}$ and let $g(X) = f(X)^2$. Show that the polynomial g has no approximate roots.
(2) Let $f \in \mathbb{Z}_{(p)}[X]$ and assume that f has no repeated roots in $\mathbb{C}$. Show that if $f(X) \equiv 0 \bmod p^N$ has solutions for all N, then f has an approximate root.

4.2 QUADRATIC CONGRUENCES

In this section, we'll consider congruences of the form

$$x^2 \equiv d \mod n.$$

In the case that n is a prime number, we saw in Section 3.5 a fast method for determining whether such a congruence has solutions. In this section, we'll extend the method to all positive integers n.

For example, consider the congruence

$$x^2 \equiv 5 \mod \left(11^{20} \cdot 19^{50} \cdot 199^{40}\right). \tag{4.5}$$

We'll find out whether this congruence has any solutions. The first step is to break the problem down into congruences modulo powers of primes. This step uses the following corollary to the Chinese Remainder Theorem.

Corollary 4.1. *Let f be a polynomial over $\mathbb{Z}$ and let n and m be coprime positive integers. The congruence $f(x) \equiv 0 \bmod nm$ has a solution if and only if both $f(x) \equiv 0 \bmod n$ and $f(x) \equiv 0 \bmod m$ have solutions.*

Proof. If $f(x) \equiv 0 \bmod nm$, then $f(x) \equiv 0 \bmod n$ and $f(x) \equiv 0 \bmod m$.

Conversely, suppose x_1 is a solution modulo n and x_2 is a solution modulo m. By the existence part of the Chinese Remainder Theorem, there exists an integer x such that $x \equiv x_1 \bmod n$ and $x \equiv x_2 \bmod m$. This implies $f(x) \equiv 0 \bmod n$ and $f(x) \equiv 0 \bmod m$. By the uniqueness part of the Chinese Remainder Theorem, we must have $f(x) \equiv 0 \bmod nm$. □

Using the corollary, we see that Equation (4.5) has a solution if and only if each of the following three congruences has a solution:

$$x^2 \equiv 5 \bmod 11^{20}, \quad x^2 \equiv 5 \bmod 19^{50}, \quad x^2 \equiv 5 \bmod 199^{40}.$$

We'll consider each of these three congruences using Hensel's Lemma. By trial and error, we can spot that $4^2 \equiv 5 \bmod 11$. This means that $a_0 = 4$ is a root of the polynomial $f(X) = X^2 - 5$ modulo the prime 11. We'll check that $a_0 = 4$ satisfies the conditions of Hensel's Lemma:

$$f(a_0) \equiv 0 \mod 11^{2c+1}, \quad \text{where } c = v_{11}(f'(a_0)). \tag{4.6}$$

Since $f'(X) = 2X$, we have

$$c = v_{11}(8) = 0,$$

and therefore $11^{2c+1} = 11$ so a_0 satisfies Equation (4.6). If we let a_n be the sequence defined by the Newton–Raphson formula, then $f(a_n) \equiv 0 \bmod 11^{2^n}$. Since $2^5 \geq 20$, the congruence $f(a_5) \equiv 0 \bmod 11^{2^5}$ implies $f(a_5) \equiv 0 \bmod 11^{20}$.

A similar trial and error approach will quickly show that $10^2 \equiv 5 \bmod 19$. We may check that $a_0 = 10$ satisfies the conditions of Hensel's Lemma with $p = 19$. We therefore have solutions to the congruence modulo all powers of 19.

The prime number 199 is a bit too large to deal with by a trial and error approach. However, we can check by the Quadratic Reciprocity Law whether or not 5 is a square modulo 199. To do this, we calculate the quadratic residue symbol as follows:

$$\left(\frac{5}{199}\right) = \left(\frac{199}{5}\right) = \left(\frac{4}{5}\right) = 1.$$

This tells us that there is a solution a_0 to the congruence $a_0^2 = 5 \bmod 199$. The solution a_0 is clearly not a multiple of 199, and so we have

$$v_{199}(f'(a_0)) = v_{199}(2a_0) = 0.$$

As before, this shows that a_0 satisfies the conditions of Hensel's Lemma with $p = 199$, and so there exist solutions to our congruence modulo all powers of 199. Putting all of this together, we deduce that Equation (4.5) has solutions. In fact, the `sage` command `solve_mod(x^2-5, 11^20 * 19^50 * 199^40)` will find all eight solutions in just a few seconds.

The method which we used can be generalized as follows.

Proposition 4.2. *Let p be an odd prime and let d be an integer coprime to p. If the congruence $x^2 \equiv d$ has solutions modulo p, then it has solutions modulo p^n for every n.*

Proof. Suppose a_0 is a solution to the congruence modulo p. Let $f(X) = X^2 - d$. We have $f(a_0) \equiv 0 \bmod p$. Furthermore, $a_0 \not\equiv 0 \bmod p$, so $f'(a_0) = 2a_0 \not\equiv 0 \bmod p$. Hence, the conditions of Hensel's Lemma are satisfied with $c = 0$, and we may define recursively a sequence a_n such that $f(a_n) \equiv 0 \bmod p^{2^n}$. □

Proposition 4.2 would be false if we allowed the prime p to be 2. For example, the integer 3 is a square modulo 2, but 3 is not a square modulo 4. Furthermore, 5 is a square modulo 4, but is not a square modulo 8. In fact, all odd squares are congruent to 1 modulo 8:

$$1^2 \equiv 3^2 \equiv 5^2 \equiv 7^2 \equiv 1 \mod 8.$$

However, it turns out that if d is an odd square modulo 8, then d is also a square modulo higher powers of 2:

Proposition 4.3. *Let d be an odd integer. If the congruence $x^2 \equiv d$ has solutions modulo 8, then it has solutions modulo 2^n for every $n > 0$. This is the case if and only if $d \equiv 1 \bmod 8$.*

Proof. Suppose $d \equiv 1 \bmod 8$ and let $f(X) = X^2 - d$. We'll show that the number $a_0 = 1$ satisfies the conditions of Hensel's Lemma. We have $c = v_2(f'(1)) = v_2(2) = 1$, so $2^{2c+1} = 8$. We must therefore check that $f(1) \equiv 0 \bmod 8$. However, this follows from our assumption on d. Therefore, $a_0 = 1$ satisfies the conditions of Hensel's Lemma and we may lift it to a root of f modulo 2^n for any n.

Conversely if d is a square modulo 2^n for all n, then clearly d is a square modulo 8. We've already seen that this implies $d \equiv 1 \bmod 8$. □

We can now answer questions of the form:

for which integers n does the congruence $x^2 \equiv d \bmod n$ have solutions?

Suppose $n = \prod p_i^{m_i}$ with p_i distinct primes. The congruence has solutions modulo n if and only if it has solutions modulo $p_i^{m_i}$ for every i. However, Propositions 4.2 and 4.3 tell us when this will happen, at least when p_i is not a factor of d. In practice, the other cases can be worked out by hand.

Example. We'll find out for which integers n the congruence

$$x^2 \equiv -7 \mod n \tag{4.7}$$

has solutions. If n factorizes into primes as follows

$$n = p_1^{m_1} \cdots p_r^{m_r},$$

then by the Chinese Remainder Theorem there are solutions modulo n if and only if there are solutions modulo $p_i^{m_i}$ for each i.

If p is a prime not equal to 2 or 7, then by Proposition 4.2 the congruence has solutions modulo p^m if and only if $\left(\frac{-7}{p}\right) = 1$. We calculate this as follows:

$$\left(\frac{-7}{p}\right) = \left(\frac{-1}{p}\right)\left(\frac{7}{p}\right).$$

By the First Nebensatz and the Quadratic Reciprocity Law, we have

$$\left(\frac{-7}{p}\right) = (-1)^{\frac{p-1}{2}}(-1)^{\frac{(7-1)(p-1)}{4}}\left(\frac{p}{7}\right).$$

The signs $(-1)^{\frac{p-1}{2}}$ and $(-1)^{\frac{(7-1)(p-1)}{4}}$ are the same, so cancel each other out. This leaves us with

$$\left(\frac{-7}{p}\right) = \left(\frac{p}{7}\right).$$

The right-hand side depends only on the congruency class of p modulo 7. The quadratic residues modulo 7 are 1, 2 and 4, so we have

$$\left(\frac{-7}{p}\right) = \begin{cases} 1 & \text{if } p \equiv 1, 2, 4 \bmod 7, \\ -1 & \text{if } p \equiv 3, 5, 6 \bmod 7. \end{cases}$$

We've shown that for primes other than 2 and 7, the congruence in Equation (4.7) has solutions modulo p^m if and only if p is congruent to 1, 2 or 4 modulo 7.

It remains to consider powers of 2 and of 7. Note that $-7 \equiv 1 \bmod 8$ so by Proposition 4.3 the congruence has solutions modulo 2^m for all m. Next consider $n = 7^m$. The congruence has solutions modulo 7; indeed $x = 0$ is a solution modulo 7. However, there are no solutions modulo 7^2. To see why this is true, let's suppose we have a solution to $x^2 \equiv -7 \bmod 49$. This implies $x^2 \equiv 0 \bmod 7$, and by Euclid's Lemma x must be a multiple of 7. Therefore, x^2 is a multiple of 49, which contradicts the assumption $x^2 \equiv -7 \bmod 49$. As there are no solutions modulo 7^2, it follows that there are no solutions modulo 7^m for any $m \geq 2$.

Putting all of this together, we find that Equation (4.7) has solutions modulo n if and only if $n = ab$, where $a \in \{1, 7\}$ and all the prime factors of b are congruent to 1, 2 or 4 modulo 7.

Exercise 4.8. Which of the following congruences have solutions?

$$\begin{aligned} &x^2 \equiv 87 \quad \bmod 2, & &x^2 \equiv 87 \quad \bmod 2^{100}, \\ &x^2 \equiv 87 \quad \bmod 127, & &x^2 \equiv 87 \quad \bmod 2 \cdot 3 \cdot 127^{2000}, \\ &x^2 \equiv 87 \quad \bmod 9 \cdot 181, & &x^2 \equiv 87 \quad \bmod 29^{2000}. \end{aligned}$$

Exercise 4.9. For which n does $x^2 \equiv 5 \bmod n$ have solutions?

Exercise 4.10. For which n does $x^2 + x - 4 \equiv 0 \bmod n$ have solutions?

Exercise 4.11. Let p be a prime number and let $f(X) = X^2 - p$. Show that the congruence $f(x) \equiv 0 \bmod p$ has a solution, but for $n \geq 2$ the congruence $f(x) \equiv 0 \bmod p^n$ does not have a solution.

Exercise 4.12. Show that for all positive integers n, the following congruence has solutions:

$$(x^2 - 2)(x^2 + 7)(x^2 + 14) \equiv 0 \mod n.$$

Exercise 4.13. Let a be a non-zero integer. Show that the following are equivalent:

(1) a is a square modulo 2^n for every n.
(2) $a = 4^m b$, where $b \equiv 1 \bmod 8$.

Exercise 4.14. Let p be a prime number and assume $p \neq 3$, and let a be an integer which is not a multiple of p. Show that if $x^3 \equiv a \bmod p$ has solutions, then $x^3 \equiv a \bmod p^n$ has solutions for all n.

4.3 p-ADIC CONVERGENCE OF SERIES

Recall that for a prime number p, the local ring $\mathbb{Z}_{(p)}$ is defined by

$$\mathbb{Z}_{(p)} = \{x \in \mathbb{Q} : v_p(x) \geq 0\}.$$

The elements of $\mathbb{Z}_{(p)}$ are the rational numbers which represent congruency classes modulo powers of p. Suppose we have a series of elements of $\mathbb{Z}_{(p)}$:

$$\sum_{n=1}^{\infty} x_n. \tag{4.8}$$

We'll describe a sense in which such a series might "converge"; this is *not* the usual convergence of series from analysis.

Definition. The series in Equation (4.8) is said to *converge p-adically* if for every r, there are only finitely many terms x_n for which $x_n \not\equiv 0 \bmod p^r$. If the series converges p-adically, then we can add up the terms in $\mathbb{Z}/p^r$ since only a finite number of these terms are non-zero modulo p^r. This means that a p-adically convergent series represents congruency classes modulo p^r for each r.

The idea of p-adic convergence turns out to be useful. Before describing the theory, we'll look at an example. Consider the binomial series expansion of $(1+3x)^{\frac{1}{2}}$.

$$(1+3x)^{\frac{1}{2}} = 1+\frac{1}{2}(3x)+\frac{(\frac{1}{2})(-\frac{1}{2})}{2!}(3x)^2+\frac{(\frac{1}{2})(-\frac{1}{2})(-\frac{3}{2})}{3!}(3x)^3+\cdots. \qquad (4.9)$$

If x is a real number, then the series converges (in the usual sense) as long as $|x| \leq \frac{1}{3}$, and it converges to the positive square root of $1+3x$. However, we'll be interested in this series in the case that x is in $\mathbb{Z}_{(3)}$. It turns out that the series converges 3-adically for all $x \in \mathbb{Z}_{(3)}$. This means that it converges in $\mathbb{Z}/3$ and in $\mathbb{Z}/9$ and in $\mathbb{Z}/27$, etc. For example, all terms in Equation (4.9) apart from the constant term are multiples of 3, and so we have

$$(1+3x)^{\frac{1}{2}} \equiv 1 \mod 3.$$

Modulo 9, all but the first two terms are zero and we have

$$(1+3x)^{\frac{1}{2}} \equiv 1+\frac{3x}{2} \mod 9.$$

Similarly, modulo 27 and modulo 81 we have:

$$\begin{aligned}(1+3x)^{\frac{1}{2}} &\equiv 1+\frac{3x}{2}-\frac{9x^2}{8} && \mod 27,\\ &\equiv 1+\frac{3x}{2}-\frac{9x^2}{8}+\frac{27x^3}{16} && \mod 81.\end{aligned}$$

The coefficients of these polynomials are in $\mathbb{Z}_{(3)}$, so we may reduce them modulo powers of 3 to obtain integer coefficients. This is much easier than it looks; for example, since $8^{-1} \equiv -1 \mod 9$, it follows that $9 \cdot 8^{-1} \equiv -9 \mod 81$. Similarly, since $16^{-1} \equiv 1 \mod 3$ we have $27 \cdot 16^{-1} \equiv 27 \mod 81$. This quickly gives us the following expansions:

$$\begin{aligned}(1+3x)^{\frac{1}{2}} &\equiv 1+6x && \mod 9,\\ &\equiv 1+15x+9x^2 && \mod 27,\\ &\equiv 1+42x+9x^2+27x^3 && \mod 81.\end{aligned}$$

The important point is that these polynomials play the same role in number theory that the power series $(1+3x)^{\frac{1}{2}}$ plays in analysis, i.e. they are square roots of $1+3x$. As an example, we'll expand out the polynomial

$(1+15x+9x^2)^2$ modulo 27:

$$\begin{aligned}\left(1+15x+9x^2\right)^2 &= 1+(2\cdot 15)x+(9+9+15^2)x^2+(2\cdot 15\cdot 9)x^3+81x^4\\ &\equiv 1+3x \mod 27.\end{aligned}$$

This means that in order to find a square root of 7 modulo 81, we can substitute $x=2$ into the power series:

$$\begin{aligned}7^{\frac{1}{2}} &= \left(1+3\cdot 2\right)^{\frac{1}{2}}\\ &\equiv 1+42\cdot 2+9\cdot 2^2+27\cdot 2^3 \mod 81\\ &\equiv 13 \qquad\qquad\qquad\qquad\qquad \mod 81.\end{aligned}$$

One can easily check that this is correct: $13^2 = 169 \equiv 7 \bmod 81$. The rest of this section will be an explanation of why this method works.

The Power Series Trick. In the example above, we saw how to find a square root of 7 modulo a power of 3 using the power series $(1+3x)^{\frac{1}{2}}$. This method is justified by a result called the Power Series Trick, which we shall prove below. The idea of the power series trick, is that whatever is true for power series of real or complex numbers, will also be true modulo p^n for power series which converge p-adically.

Before we can state the power series trick, we need some more notation. We shall write $\mathbb{Z}_{(p)}[[X]]$ for the set of all power series with coefficients in $\mathbb{Z}_{(p)}$:

$$\sum_{n=0}^{\infty} a_n X^n \quad a_n \in \mathbb{Z}_{(p)}.$$

The difference between a power series and a polynomial, is that a power series may have infinitely many terms. Note that $\mathbb{Z}_{(p)}[[X]]$ consists of all power series, whether they converge or not. We can add, subtract and multiply power series in the same way as we do polynomials, so in fact $\mathbb{Z}_{(p)}[[X]]$ is a ring. There is another operation on power series: we can sometimes compose two power series f and g to form a new power series $f\circ g$. For example, if $f(X) = \sum_{n=0}^{\infty} X^n$ and $g(X) = X + X^2$, then the power series $f\circ g$ is defined by substituting $g(X)$ into f:

$$\begin{aligned}(f\circ g)(X) &= 1+(X+X^2)+(X+X^2)^2+(X+X^2)^3+\cdots\\ &= 1+X+2X^2+3X^3+5X^4+\cdots.\end{aligned}$$

(The coefficients of $f\circ g$ in this case are the Fibonacci sequence; see Exercise 4.20.) It will turn out that $f\circ g$ is a well-defined power series as long as either f is a polynomial or g has no constant term. To understand why $f \circ g$ is not always defined, we consider another example. If we let $f(X) = \sum_{n=0}^{\infty} X^n$, then $f(1+X)$ makes no sense as a power series; for example, its constant term would be the infinite sum $1+1+1+1+\cdots$.

Suppose that f and g are power series, and that g has constant term zero:

$$f(X) = \sum_{n=0}^{\infty} a_n X^n, \quad g(X) = \sum_{m=1}^{\infty} b_m X^m.$$

To understand what we mean by the composition $f \circ g$, we naively substitute g into f and calculate:

$$\begin{aligned} f(g(X)) &= \sum_{n=0}^{\infty} a_n \left(\sum_{m=1}^{\infty} b_m X^m \right)^n \\ &= \sum_{n=0}^{\infty} a_n \left(\sum_{m_1=1}^{\infty} b_{m_1} X^{m_1} \right) \left(\sum_{m_2=1}^{\infty} b_{m_2} X^{m_2} \right) \cdots \left(\sum_{m_n=1}^{\infty} b_{m_n} X^{m_n} \right) \\ &= \sum_{d=0}^{\infty} c_d \cdot X^d, \end{aligned}$$

where the coefficients c_d are defined by:

$$c_d = \sum_{n=1}^{d} a_n \left(\sum_{\substack{m_1,\ldots,m_n \in \{1,\ldots,d\}, \\ \text{such that} \\ m_1+m_2+\cdots+m_n=d}} b_{m_1} \cdots b_{m_n} \right). \tag{4.10}$$

We shall regard Equation (4.10) as the definition of the composition $f \circ g$. The formula for c_d is very cumbersome and we will not want to use it directly for calculations. However, there are a few important facts which follow easily from this definition:

(1) The sum defining c_d is a finite sum, so c_d is another element of the ring $\mathbb{Z}_{(p)}$. In particular, this shows that $f \circ g$ is another element of $\mathbb{Z}_{(p)}[[X]]$.
(2) The sum defining c_d depends only on the coefficients a_n and b_n of f and g for $n \leq d$. This means that if we know $f(X)$ and $g(X)$ up to $O(X^d)$, then we can calculate $f(g(X))$ up to $O(X^d)$.
(3) The congruency class of c_d modulo p^r depends only on the congruency classes of a_n and b_n modulo p^r. This means that if we write $\bar{f}$ for the

image of f in $(\mathbb{Z}/p^r)[[X]]$, then we have

$$\overline{f \circ g} = \bar{f} \circ \bar{g}.$$

Using these facts, we can now prove the main result of this section.

Lemma 4.2 (The Power Series Trick). *Let f, g and h be power series with coefficients in $\mathbb{Z}_{(p)}$ and assume either that g has zero constant term, or that f is a polynomial (so that $f \circ g$ is a well-defined power series). Suppose that*

(i) *$f(x)$, $g(x)$ and $h(x)$ converge for sufficiently small real numbers x, and for such numbers we have $f(g(x)) = h(x)$;*
(ii) *$f(x)$, $g(x)$ and $h(x)$ converge p-adically for all $x \in \mathbb{Z}_{(p)}$.*

Then $h = f \circ g$ as power series, and for all $x \in \mathbb{Z}_{(p)}$ we have

$$f(g(x)) \equiv h(x) \mod p^n.$$

Proof. We'll first show that $h = f \circ g$ as power series. This part of the proof is calculus, and has no number theoretical content.

Assume we are in the case that g has constant term 0; the other case will be discussed later. Fix a positive integer n and let f_n and g_n be the polynomials obtained from f and g by truncating after the X^n term. This implies

$$|g(x) - g_n(x)| \ll |x|^{n+1}.$$

Here we are using the notation $A(x) \ll B(x)$ or $A(x) = O(B(x))$, to mean that there is a constant c, such that for all x in a small open interval I around 0 we have $A(x) \leq c \cdot B(x)$. We'll regard n as fixed, so the implied constant c and the interval I are allowed to depend on n. We'll also write $(f \circ g)_n$ for the first n terms of the formal power series $f \circ g$. To show that $h = f \circ g$, we must show that $h(x) = (f \circ g)_n(x) + O(|x|^{n+1})$.

To make this more precise, we let J be an open interval containing 0, small enough so that $f(y)$ converges and $f'(y)$ is bounded for $y \in J$. We also choose an open interval I containing 0, small enough so that $g(x)$ and $h(x)$ both converge and both $g(x)$ and $g_n(x)$ are in J for all $x \in I$; furthermore, we assume $h(x) = f(g(x))$ for $x \in I$.

In what follows we let x be in I. We are trying to bound $h(x) - (f \circ g)_n(x)$. As a first step, note that by the triangle inequality, we have

$$\begin{aligned}\big|h(x)-(f\circ g)_n(x)\big| \le \big|f(g(x))-f(g_n(x))\big| \\ +\big|f(g_n(x))-f_n(g_n(x))\big| \\ +\big|f_n(g_n(x))-(f\circ g)_n(x)\big|.\end{aligned} \tag{4.11}$$

Since f' is bounded on the interval J which contains both $g(x)$ and $g_n(x)$, we have (by the Mean Value Theorem):

$$\big|f(g(x))-f(g_n(x))\big| \ll |g(x)-g_n(x)| \ll |x|^{n+1}. \tag{4.12}$$

Since g has constant term 0, we have $|g_n(x)| \ll |x|$, and this implies:

$$\big|f(g_n(x))-f_n(g_n(x))\big| \ll |g_n(x)|^{n+1} \ll |x|^{n+1}. \tag{4.13}$$

Finally, we observed in Equation (4.10) that the first n terms of $f\circ g$ may be expressed in terms of the first n terms of f and g. These terms are therefore exactly the same as the first n terms of the polynomial $f_n(g_n(X))$. This implics

$$\big|f_n(g_n(x))-(f\circ g)_n(x)\big| \ll |x|^{n+1}. \tag{4.14}$$

Substituting Equations (4.12)–(4.14) into Equation (4.11), we get

$$h(x) = (f\circ g)_n(x) + O(|x|^{n+1}).$$

This shows that the first n coefficients of h are the same as those of $f\circ g$. Repeating the argument with larger and larger n, we find that $h = f\circ g$.

We'll now consider the case where g has a non-zero constant term but f is a polynomial. If we define $G(X) = g(X) - g(0)$ and $F(X) = f(X + g(0))$, then we have $F(G(x)) = h(x)$ for small real numbers x. Furthermore, $G(X)$ has zero constant term, so by the argument given above we know that $h = F\circ G$ as power series. This implies $h = f\circ g$ in this case as well.

It remains to prove the congruence modulo p^n. We'll write $\bar{f}$, $\bar{g}$ and $\bar{h}$ for the images of f, g and h in the power series ring $(\mathbb{Z}/p^n)[[X]]$, and we'll write $\bar{x}$ for the image in $\mathbb{Z}/p^n$ of an element $x \in \mathbb{Z}_{(p)}$. As $f(1)$, $g(1)$ and $h(1)$ all converge p-adically, it follows that f, g and h have only finitely many non-zero coefficients modulo p^n. Therefore, the reductions $\bar{f}$, $\bar{g}$ and $\bar{h}$ are polynomials over $\mathbb{Z}/p^n$. Since $h = f\circ g$, it follows that $\bar{h} = \bar{f}\circ\bar{g}$, and the composition $\bar{f}\circ\bar{g}$ is the usual composition of polynomials. This implies $\bar{h}(\bar{x}) \equiv \bar{f}(\bar{g}(\bar{x})) \mod p^n$. Equivalently, $h(x) \equiv f(g(x)) \bmod p^n$. □

Checking for p-adic convergence. The Power Series Trick is a useful way of proving congruence relations. However, in order to use this result, we need a way of checking whether a power series converges p-adically.

Lemma 4.3. *A series* $\sum x_n$ *of elements in* $\mathbb{Z}_{(p)}$ *converges p-adically if and only if* $v_p(x_n)$ *tends to infinity as* $n \to \infty$.

Proof. If $v_p(x_n) \to \infty$, then for large values of n, $v_p(x_n) \geq r$, which is equivalent to saying that $x_n \equiv 0 \bmod p^r$. □

Many power series have the number $n!$ in their coefficients, and so the following lemma will help to check for p-adic convergence. In the lemma, we use the notation $\lfloor x \rfloor$ to mean the *floor* of a real number x, i.e. the largest integer which is less than or equal to x.

Lemma 4.4. *For any positive integer* n, $v_p(n!) = \left\lfloor \frac{n}{p} \right\rfloor + \left\lfloor \frac{n}{p^2} \right\rfloor + \left\lfloor \frac{n}{p^3} \right\rfloor + \cdots$. *Furthermore, we have* $v_p(n!) \leq \dfrac{n}{p-1}$.

For a fixed n, only finitely many of the integers $\left\lfloor \frac{n}{p^a} \right\rfloor$ are non-zero, so the sum in the lemma is actually finite. For example, we can calculate $v_3(100!)$ as follows.

$$v_3(100!) = \left\lfloor \frac{100}{3} \right\rfloor + \left\lfloor \frac{100}{9} \right\rfloor + \left\lfloor \frac{100}{27} \right\rfloor + \left\lfloor \frac{100}{81} \right\rfloor = 33 + 11 + 3 + 1 = 48.$$

The upper bound in the second half of the lemma is $v_3(100!) \leq \frac{100}{2} = 50$, which is quite close to the exact value of 48 (see Exercise 4.19).

Proof of Lemma 4.4. We begin with the equation

$$v_p(n!) = \sum_{i=1}^{n} v_p(i).$$

If we let $s(r)$ be the number of terms in this sum which are equal to r, then we can rewrite the equation as

$$v_p(n!) = 1 \cdot s(1) + 2 \cdot s(2) + 3 \cdot s(3) + \cdots .$$

By definition, $s(r)$ is the number of integers i between 1 and n, such that $v_p(i) = r$. There are $\lfloor \frac{n}{p^r} \rfloor$ values of i which are multiples of p^r; however $\lfloor \frac{n}{p^{r+1}} \rfloor$ of these values are also multiples of p^{r+1}. From this, we see that $s(r) = \lfloor \frac{n}{p^r} \rfloor - \lfloor \frac{n}{p^{r+1}} \rfloor$. Substituting this into our formula for $v_p(n!)$, we get

$$\begin{aligned} v_p(n!) &= \left(\left\lfloor \frac{n}{p} \right\rfloor - \left\lfloor \frac{n}{p^2} \right\rfloor \right) + 2 \left(\left\lfloor \frac{n}{p^2} \right\rfloor - \left\lfloor \frac{n}{p^3} \right\rfloor \right) + 3 \left(\left\lfloor \frac{n}{p^3} \right\rfloor - \left\lfloor \frac{n}{p^4} \right\rfloor \right) + \cdots \\ &= \left\lfloor \frac{n}{p} \right\rfloor + (2-1) \left\lfloor \frac{n}{p^2} \right\rfloor + (3-2) \left\lfloor \frac{n}{p^3} \right\rfloor + \cdots \\ &= \left\lfloor \frac{n}{p} \right\rfloor + \left\lfloor \frac{n}{p^2} \right\rfloor + \left\lfloor \frac{n}{p^3} \right\rfloor + \cdots . \end{aligned}$$

This proves the first equation of the lemma. To prove the inequality, note that $\lfloor x \rfloor \leq x$, and therefore

$$\sum_{i=1}^{\infty} \left\lfloor \frac{n}{p^i} \right\rfloor \leq \sum_{i=1}^{\infty} \frac{n}{p^i}.$$

The right-hand side is a geometric progression, and is equal to $\frac{n}{p-1}$. □

We now return to the example which we considered earlier. The binomial expansion of $(1+3x)^{\frac{1}{2}}$ is given by

$$\sum_{n=0}^{\infty} \frac{\left(\frac{1}{2}\right)\left(-\frac{1}{2}\right)\cdots\left(\frac{3-2n}{2}\right)}{n!}(3x)^n.$$

We'll see that this power series converges 3-adically for all $x \in \mathbb{Z}_{(3)}$. To prove this, we must show that the 3-adic valuations of the terms tend to infinity. Since $x \in \mathbb{Z}_{(3)}$, we have $v_3(x) \geq 0$. Similarly, the numbers $\frac{1}{2}, -\frac{1}{2}, -\frac{3}{2}, \ldots, \frac{3-2n}{2}$ are all in $\mathbb{Z}_{(3)}$ so have valuation ≥ 0. From this, we see that

$$v_3\left(\frac{\left(\frac{1}{2}\right)\left(-\frac{1}{2}\right)\cdots\left(\frac{3-2n}{2}\right)}{n!}(3x)^n\right) \geq v_3\left(\frac{3^n}{n!}\right) = n - v_3(n!).$$

Using the bound in Lemma 4.4, we have

$$v_3\left(\frac{\left(\frac{1}{2}\right)\left(-\frac{1}{2}\right)\cdots\left(\frac{3-2n}{2}\right)}{n!}(3x)^n\right) \geq n - \frac{n}{2} = \frac{n}{2}.$$

In particular, the valuations tend to infinity, so by Lemma 4.3 the series converges 3-adically. We can now use the Power Series Trick to explain why our series gives us square roots of $1+3x$. Let

$$f(X) = X^2, \quad g(X) = (1+3X)^{\frac{1}{2}}, \quad h(X) = 1+3X.$$

For small real numbers x, we certainly have $f(g(x)) = h(x)$. Applying the Power Series Trick, we have the following congruence for all $x \in \mathbb{Z}_{(3)}$:

$$\left((1+3x)^{\frac{1}{2}}\right)^2 \equiv 1+3x \mod 3^n.$$

In particular, taking $x = 2$ and $n = 4$, we found

$$13^2 \equiv 7 \mod 81.$$

Exercise 4.15. Calculate $v_2(120!)$, $v_3\left(\frac{100!}{20!80!}\right)$ and $v_{11}((11^5)!)$.

Exercise 4.16. Let p be a prime number. Which of the following series converge p-adically?

$$\sum_{n=0}^{\infty} p^n, \quad \sum_{n=0}^{\infty} (p^n+1), \quad \sum_{n=0}^{\infty} n!, \quad \sum_{n=0}^{\infty} n, \quad \sum_{n=0}^{\infty} \frac{(2n)!}{n!}, \quad \sum_{n=0}^{\infty} \frac{(2n)!}{(n!)^2}.$$

Exercise 4.17. Suppose the series $\sum a_n$ and $\sum b_n$ converge p-adically. Prove that $\sum(a_n + b_n)$ converges p-adically, and show that for all r,

$$\sum(a_n + b_n) \equiv \sum a_n + \sum b_n \mod p^r.$$

Let $c_n = a_0 b_n + a_1 b_{n-1} + \cdots + a_n b_0$. Show that $\sum c_n$ converges p-adically and

$$\sum c_n \equiv \left(\sum a_n\right)\left(\sum b_n\right) \mod p^r.$$

Exercise 4.18. Show that the following holds for all $x \in \mathbb{Z}_{(p)}$ and all r:

$$\sum_{n=0}^{\infty}(px)^n \equiv \frac{1}{1-px} \mod p^r.$$

Hence, find $11^{-1} \bmod 5^4$ without using Euclid's algorithm.

Exercise 4.19. Show that $\frac{n}{p-1} - v_p(n!) < \log_p(n) + \frac{p}{p-1}$, where $\log_p$ is the logarithm to the base p.

Exercise 4.20. Let $f(X) - \sum_{n=0}^{\infty} X^n$. Show that $f(X+X^2) = \sum F_n \cdot X^n$, where the numbers F_n are the Fibonacci sequence, defined by $F_0 = F_1 = 1$ and $F_{n+2} = F_n + F_{n+1}$.

4.4 p-ADIC LOGARITHMS AND EXPONENTIAL MAPS

Throughout this section, we let p be an odd prime number. We'll write $p\mathbb{Z}/p^n$ for the set of elements in $\mathbb{Z}/p^n$ which are multiples of p. For example,

$$3\mathbb{Z}/27 = \{0, 3, 6, 9, 12, 15, 18, 21, 24\}.$$

This set is obviously closed under addition, so it is an additive subgroup of $\mathbb{Z}/p^n$ with p^{n-1} elements.

Similarly, we write $1 + p\mathbb{Z}/p^n$ for the set of elements of the form $1 + px$ for $x \in \mathbb{Z}/p^n$. For example,

$$1 + 3\mathbb{Z}/27 = \{1, 4, 7, 10, 13, 16, 19, 22, 25\}.$$

If x and y are both congruent to 1 modulo p, then so is xy, so $1 + p\mathbb{Z}/p^n$ is a multiplicative subgroup of $(\mathbb{Z}/p^n)^\times$, also with p^{n-1} elements.

At first sight, there's no reason to think that these two groups have the same structure, since the operations $+$ and $\times$ are very different. However, we will show in Theorem 4.2 that these groups are isomorphic. The isomorphism $p\mathbb{Z}/p^n \to 1 + p\mathbb{Z}/p^n$ is the map $px \mapsto \exp(px)$ given by the exponential power series. Its inverse is given by the power series for $\log(1+px)$. The theorem will be proved using the Power Series Trick from the previous section.

We'll write $\exp(X)$, for the formal power series

$$\exp(X) = 1 + X + \frac{X^2}{2!} + \frac{X^3}{3!} + \cdots .$$

For real numbers x, the series $\exp(x)$ converges and we have $\exp(x) = e^x$. If we write log for the logarithm to the base e, then for all real x we have $\log(\exp(x)) = x$, and for all positive x we have $\exp(\log(x)) = x$. When we write $\log(1+X)$, we'll mean the following power series:

$$\log(1+X) = X - \frac{X^2}{2} + \frac{X^3}{3} - \frac{X^4}{4} + \cdots .$$

This makes sense because of the following:

Lemma 4.5. *For real numbers x with $|x| < 1$, we have*

$$\log(1+x) = x - \frac{x^2}{2} + \frac{x^3}{3} - \frac{x^4}{4} + \cdots .$$

Proof. If $|x| < 1$, then the formula for a geometric progression gives

$$\frac{1}{1+x} = 1 - x + x^2 - x^3 + \cdots .$$

Integrating this equation term by term, we get the formula of the lemma. □

Theorem 4.2. *Let p be an odd prime number. Then the power series $\log(1+px)$ and $\exp(px)$ converge p-adically for all $x \in \mathbb{Z}_{(p)}$. They give inverse functions*

$$\log : 1 + p\mathbb{Z}/p^n \to p\mathbb{Z}/p^n, \quad \exp : p\mathbb{Z}/p^n \to 1 + p\mathbb{Z}/p^n.$$

These functions are group isomorphisms between the multiplicative group $1 + p\mathbb{Z}/p^n$ and the additive group $p\mathbb{Z}/p^n$.

Before proving the theorem, we'll show how it can be used for calculations. As a toy example, we'll calculate these power series explicitly modulo 3^3. We have

$$\exp(3x) \equiv 1 + 3x + \frac{9x^2}{2} + \frac{27x^3}{6} \mod 27.$$

Since $2^{-1} \equiv 2 \bmod 3$, it follows that $\frac{9}{2} \equiv 18 \bmod 27$, so we can reduce this to

$$\exp(3x) \equiv 1 + 3x + 18x^2 + 18x^3 \mod 27.$$

Similarly, we have

$$\begin{aligned} \log(1+3x) &\equiv 3x - \frac{9x^2}{2} + \frac{27x^3}{3} && \mod 27 \\ &\equiv 3x + 9x^2 + 9x^3 && \mod 27. \end{aligned}$$

The theorem says that these power series give inverse functions. We can check this directly by substituting one into the other:

$$\begin{aligned} \log(\exp(3x)) &\equiv \log(1 + 3x + 18x^2 + 18x^3) \mod 27 \\ &\equiv \log(1+3y) \mod 27, \qquad \text{where } y = x + 6x^2 + 6x^3. \end{aligned}$$

Now using our formula for $\log(1+3x)$, we have

$$\log(\exp(3x)) \equiv 3y + 9y^2 + 9y^3 \mod 27.$$

Since $y = x \bmod 3$, it follows that $9y^2 \equiv 9x^2 \bmod 27$ and $9y^3 \equiv 9x^3 \bmod 27$. As predicted, this implies

$$\log(\exp(3x)) \equiv 3(x + 6x^2 + 6x^3) + 9x^2 + 9x^3 \equiv 3x \mod 27.$$

A similar calculation shows that $\exp(\log(1+3x)) \equiv 1 + 3x \bmod 27$.

Calculating discrete logarithms. As an example of what can be done with the theorem, we'll solve the congruence

$$7^x \equiv 13 \mod 27. \tag{4.15}$$

Since 7 and 13 are both in the group $1 + 3\mathbb{Z}/27$, we can take logarithms of both sides of the equation. Using the fact that log is a homomorphism, we see that $\log(7^x) \equiv \log(7) \cdot x \bmod 27$. Therefore, our equation becomes

$$\log(7) \cdot x \equiv \log(13) \mod 27.$$

We've already calculated that $\log(1+3x) \equiv 3x + 9x^2 + 9x^3 \bmod 27$. Substituting $x = 2$ and $x = 4$, we find $\log(7) \equiv 6 \bmod 27$ and $\log(13) \equiv 3 \bmod 27$, respectively. Therefore, our equation reduces to

$$6x \equiv 3 \mod 27.$$

Solving this, we find $x \equiv 5 \bmod 9$. We can check that $x = 5$ is indeed a solution to Equation (4.15). Furthermore, the element 7 has order 9 in the multiplicative group, so we have found all the solutions to the congruence.

Fractional powers. If p is an odd prime and $a \equiv 1 \bmod p$, then we may define for all $x \in \mathbb{Z}_{(p)}$ a congruency class $a^x \bmod p^n$ by the formula

$$a^x \equiv \exp(x \log(a)) \mod p^n.$$

Note that we require $a \equiv 1 \bmod p$ in order that $\log(a)$ converges p-adically. With this notation, we have the usual rules for manipulating powers:

$$(ab)^x \equiv a^x b^x \bmod p^n, \quad a^{x+y} \equiv a^x a^y \bmod p^n, \quad a^{xy} \equiv (a^x)^y \bmod p^n.$$

We can prove these properties using the properties of log and exp given in Theorem 4.2 (see Exercise 4.25).

As an example, we'll calculate $4^{\frac{1}{2}} \bmod 27$. By our formula for $\log(1+3x) \bmod 27$, we have

$$\log(4) \equiv \log(1+3) \equiv -6 \mod 27.$$

Therefore, for any $x \in \mathbb{Z}_{(3)}$, we have

$$\begin{aligned} 4^x &\equiv \exp(-6x) && \mod 27 \\ &\equiv 1 - 6x + \frac{36x^2}{2} - \frac{6^3x^3}{6} && \mod 27 \\ &\equiv 1 - 6x + 18x^2 - 9x^3 && \mod 27. \end{aligned}$$

In particular,

$$4^{\frac{1}{2}} \equiv 1 - 3 + \frac{18}{4} - \frac{9}{8} \equiv -2 \mod 27.$$

Proof of Theorem 4.2. First consider the exponential series for $x \in \mathbb{Z}_{(p)}$:

$$\exp(px) = \sum_{n=0}^{\infty} \frac{(px)^n}{n!}.$$

To prove p-adic convergence, we must show that $v_p(\frac{(px)^n}{n!})$ tends to infinity as $n \to \infty$. Since x is in $\mathbb{Z}_{(p)}$, we know that $v_p(x) \geq 0$. This implies

$$v_p\left(\frac{(px)^n}{n!}\right) = n \cdot v_p(px) - v_p(n!) \geq n - v_p(n!).$$

We showed in Lemma 4.4 that $v_p(n!) \leq \frac{n}{p-1}$. Therefore,

$$v_p\left(\frac{(px)^n}{n!}\right) \geq \left(\frac{p-2}{p-1}\right) n.$$

Since p is bigger than 2, the number $\frac{p-2}{p-1}$ is positive. This shows that $v_p\left(\frac{(px)^n}{n!}\right)$ tends to infinity as $n \to \infty$. If $n > 0$ then $v_p(\frac{(px)^n}{n!}) > 0$, so all terms apart from the constant term 1 are multiples of p. This shows that $\exp(px) \equiv 1 \bmod p$. Furthermore, the following power series has coefficients in $\mathbb{Z}_{(p)}$:

$$f(X) = \frac{\exp(pX) - 1}{p}.$$

To prove convergence of $\log(1+px)$, note that the n-th term is $\pm\frac{(px)^n}{n}$. Since n is a factor of $n!$, it follows that $v_p(n) \leq v_p(n!) \leq \frac{n}{p-1}$. This implies

$$v_p\left(\pm\frac{(px)^n}{n}\right) \geq n - \frac{n}{p-1} = \left(\frac{p-2}{p-1}\right)n.$$

As with the exponential series, we deduce that $\log(1 + px)$ converges p-adically, and all of its coefficients are multiples of p. Therefore, the power series

$$g(X) = \frac{1}{p}\log(1 + pX)$$

also has coefficients in $\mathbb{Z}_{(p)}$.

We'll next show that exp and log give well-defined functions between the sets $p\mathbb{Z}/p^n$ and $1+p\mathbb{Z}/p^n$. This is not quite obvious, because the power series $\exp(X)$ and $\log(1 + X)$ do not have coefficients in $\mathbb{Z}_{(p)}$. Instead, we shall use the power series f and g which we have just defined. If $px \equiv py \bmod p^n$, then $x \equiv y \bmod p^{n-1}$; this implies $f(x) \equiv f(y) \bmod p^{n-1}$, and so $\exp(px) \equiv \exp(py) \bmod p^n$. Therefore, $\exp(px)$ is well-defined modulo p^n for $px \in p\mathbb{Z}/p^n$. Furthermore, since $\exp(px) \equiv 1 \bmod p$, it follows that $\exp(px) \in 1 + p\mathbb{Z}/p^n$. A similar argument using the power series g shows that the logarithm series gives a well-defined function from $1 + p\mathbb{Z}/p^n$ to $p\mathbb{Z}/p^n$.

We can compose the power series f and g in either order, since they both have zero constant term. For small real numbers x, we have $g(f(x)) = x$ and $f(g(x)) = x$. Hence, by the Power Series Trick, we have congruences for all $x \in \mathbb{Z}_{(p)}$:

$$f(g(x)) \equiv x \bmod p^n, \quad g(f(x)) \equiv x \bmod p^n.$$

Multiplying by p we get

$$\exp(\log(1 + px)) \equiv 1 + px \bmod p^n, \quad \log(\exp(px)) \equiv px \bmod p^n.$$

We've shown that the functions log and exp are inverses, so they are bijective.

It remains to show that exp is a group homomorphism, i.e.

$$\exp(p(x + y)) \equiv \exp(px) \cdot \exp(py) \mod p^n.$$

To prove this, we use the Power Series Trick again. For any positive integer a, we have (for all real x):

$$\exp(pax) = \exp(px)^a.$$

Hence, by the power series trick, we have for all $x \in \mathbb{Z}_{(p)}$:

$$\exp(pax) \equiv \exp(px)^a \mod p^n.$$

Setting $x = 1$ we get (for all positive integers a):

$$\exp(pa) \equiv \exp(p)^a \mod p^n.$$

From this, it obviously follows that

$$\exp(p(a+b)) \equiv \exp(p)^{a+b} \equiv \exp(pa) \cdot \exp(pb) \mod p^n.$$

As this is true for all positive integers a and b, it must be true for a and b in $\mathbb{Z}/p^n$. Therefore, exp is an isomorphism of groups. □

Exercise 4.21.

(1) Calculate $\log(1 + 5x)$ and $\exp(5x)$ modulo 5^4.
(2) Hence, find a polynomial formula for $11^a \bmod 5^4$ when a is in $\mathbb{Z}_{(5)}$.
(3) Hence, find a solution to the congruence $x^4 \equiv 11 \bmod 5^4$.

Exercise 4.22.

(1) Calculate the polynomial $\log(1 + 3x) \bmod 81$.
(2) Calculate $\log(-5)$ and $\log(16)$ modulo 81.
(3) Find an integer x such that $2^x \equiv -5 \bmod 81$.

Exercise 4.23. Show that the series $\exp(2)$ does not converge 2-adically but $\exp(4x)$ converges 2-adically for all $x \in \mathbb{Z}_{(2)}$.

Exercise 4.24.

(1) Show that the power series $\log(1 + 2x)$ converges 2-adically for all $x \in \mathbb{Z}_{(2)}$.
(2) Let $f(x) = 2x + 2x^2$. Show that for all sufficiently small real numbers x, we have

$$\log(1 + 2f(x)) = 2\log(1 + 2x).$$

(3) Using the power series trick, prove the following congruence for all N:

$$\sum_{n=1}^{\infty} \frac{2^n}{n} \equiv 0 \mod 2^N.$$

(4) The *dilogarithm* power series $\mathrm{Li}_2(X)$ is defined by

$$\mathrm{Li}_2(X) = \sum_{n=1}^{\infty} \frac{X^n}{n^2}.$$

The series $\mathrm{Li}_2(x)$ converges for real numbers x such that $|x| < 1$. Show that for sufficiently small real numbers x, we have

$$\mathrm{Li}_2(x) + \mathrm{Li}_2\left(\frac{x}{x-1}\right) = -\frac{1}{2}\log(1-x)^2.$$

(5) Hence, prove the following congruence for all N:

$$\sum_{n=1}^{\infty} \frac{2^n}{n^2} \equiv 0 \mod 2^N.$$

Exercise 4.25. Let p be an odd prime. Using properties of log and exp, prove the following for all $a, b \equiv 1 \bmod p$ and all $x, y \in \mathbb{Z}_{(p)}$:

$$a^x \equiv 1 \bmod p, \qquad a^1 \equiv a \bmod p^n, \qquad (ab)^x \equiv a^x b^x \bmod p^n,$$
$$a^{x+y} \equiv a^x a^y \bmod p^n, \qquad a^{xy} \equiv (a^x)^y \bmod p^n.$$

Exercise 4.26. (In Section 4.3, we used the binomial expansion of $(1+3x)^{\frac{1}{2}}$ to find square roots modulo powers of 3. In this section, we defined fractional powers using exp and log. This exercise shows that the two ways of defining fractional powers give the same answer.) Let p be an odd prime number and let $a \in \mathbb{Z}_{(p)}$. Show that the binomial expansion of $(1+px)^a$ converges p-adically for all $x \in \mathbb{Z}_{(p)}$. Using the Power Series Trick, show that we have an equality of power series

$$(1+pX)^a = \exp(a \cdot \log(1+pX)).$$

4.5 TEICHMÜLLER LIFTS

In this section, we again let p be an odd prime number. Recall that we have a subgroup

$$1 + p\mathbb{Z}/p^n \subset (\mathbb{Z}/p^n)^\times,$$

consisting of the elements which are congruent to 1 modulo p. We showed that this subgroup is isomorphic to the additive group $p\mathbb{Z}/p^n$, and this isomorphism is helpful in many calculations. In this section, we'll find another subgroup of $(\mathbb{Z}/p^n)^\times$, called the subgroup of *Teichmüller lifts*. It will turn out that $(\mathbb{Z}/p^n)^\times$ is the direct sum of these two subgroups.

We'll begin with the definition of a Teichmüller lift. Let x be an element of $\mathbb{Z}_{(p)}$ which is not a multiple of p, and consider the sequence

$$x, x^p, x^{p^2}, x^{p^3}, \ldots \tag{4.16}$$

Each term in the sequence is the pth power of the preceding term, because $x^{p^{n+1}} = (x^{p^n})^p$. By Fermat's Little Theorem, we have $x^p \equiv x \bmod p$. Therefore, the sequence is actually constant modulo p: every term in the sequence is congruent to x modulo p.

If we examine this sequence modulo p^2, then it is not constant. However, it becomes constant after the second term in the sequence. For example, consider the case $x = 3$, $p = 5$. The powers 3^{5^n} modulo 5^2 are

$$\begin{aligned} 3^{5^0} &= 3, \\ 3^{5^1} &= 3^5 \equiv 18 \mod 25, \\ 3^{5^2} &\equiv 18^5 \equiv 18 \mod 25, \\ 3^{5^3} &\equiv 18^5 \equiv 18 \mod 25. \end{aligned}$$

Since $18^5 \equiv 18 \bmod 25$, all terms apart from the first term are congruent to 18. If we calculate the sequence modulo 5^3 rather than modulo 5^2, then the first three terms in the sequence may all be different, but after that the sequence is constant:

$$\begin{aligned} 3^{5^0} &= 3, \\ 3^{5^1} &= 3^5 \equiv 118 \mod 125, \\ 3^{5^2} &\equiv 118^5 \equiv 68 \mod 125, \\ 3^{5^3} &\equiv 68^5 \equiv 68 \mod 125. \end{aligned}$$

We'll show in Proposition 4.4 that this happens more generally: the sequence in Equation (4.16) is eventually constant modulo any power of p. We shall define the *Teichmüller lift* of x to be the limiting value of the sequence in $(\mathbb{Z}/p^n)^\times$. A more explicit definition is the following.

Definition. Let p be an odd prime number and let $x \not\equiv 0 \bmod p$. The *Teichmüller lift* of x to $\mathbb{Z}/p^n$ is defined to be the congruency class

$$T(x) \equiv x^{p^{n-1}} \mod p^n.$$

The following lemma is useful for calculating Teichmüller lifts.

Lemma 4.6. *Let p be a prime number and let n be a positive integer. If $x \equiv y \bmod p^n$, then $x^p \equiv y^p \bmod p^{n+1}$.*

Proof. Since $x \equiv y \bmod p^n$, we have $x = y + p^n z$ for some integer z. Expanding out using the Binomial Theorem, we get

$$x^p = (y + p^n z)^p = y^p + p \cdot y^{p-1} \cdot p^n z + \cdots + (p^n z)^p.$$

All terms in this expansion, apart from the first term y^p, are multiples of p^{n+1}, so we have $x^p \equiv y^p \bmod p^{n+1}$. □

As an example of how to use the lemma, we'll calculate $T(37)$ modulo 5^3. By definition, this is the congruency class of 37^{25} modulo 125. Before using the lemma, it's worth considering whether we could calculate this power by the method of Section 3.2, i.e. by reducing the power 25 modulo $\varphi(125)$. However, this method will not help us, because $\varphi(125) = 100$, which is bigger than the power 25. Instead, to calculate $T(37)$, we begin with the congruence

$$37 \equiv 2 \mod 5.$$

Using Lemma 4.6, it follows that

$$37^5 \equiv 2^5 \mod 25.$$

This simplifies to $37^5 \equiv 7 \bmod 25$. Using Lemma 4.6 again, we get

$$37^{25} \equiv 7^5 \mod 125.$$

At this point we could calculate 7^5 directly, and then reduce the answer modulo 125. However, it's far easier to expand $(2+5)^5$ using the binomial theorem:

$$(2+5)^5 = 2^5 + 5 \cdot 2^4 \cdot 5 + 10 \cdot 2^3 \cdot 5^2 + \cdots + 5^5.$$

All terms in this expansion are multiples of 125 apart from the first two terms, so we have

$$(2+5)^5 \equiv 2^5 + 25 \cdot 2^4 \mod 125.$$

Furthermore, since $2^4 \equiv 1 \bmod 5$, it follows that $25 \cdot 2^4 \equiv 25 \bmod 125$. This gives us

$$37^{25} \equiv 32 + 25 \equiv 57 \mod 125.$$

This shows that the Teichmüller lift of 37 modulo 5^3 is 57.

The main properties of Teichmüller lifts are listed in the following result.

Proposition 4.4. *Let p be an odd prime number and let $x \in (\mathbb{Z}/p^n)^\times$.*

(1) *For $r > n-1$, we have $x^{p^r} \equiv T(x) \bmod p^n$.*
(2) *The element $T(x)$ satisfies $T(x)^{p-1} \equiv 1 \bmod p^n$.*

(3) *$T(x)$ only depends on x modulo p. Furthermore, $T(x) \equiv x \bmod p$.*
(4) *The map $T : \mathbb{F}_p^\times \to (\mathbb{Z}/p^n)^\times$ is an injective group homomorphism.*

Another way of saying the first part of the proposition, is that $T(x)$ is the limiting value of the sequence x^{p^r} in $\mathbb{Z}/p^n$.

Before proving the proposition, we shall use it to calculate all the Teichmüller lifts modulo 125. By the third part of the proposition, $T(x)$ depends only on x modulo 5, so we only need to calculate $T(x)$ for $x = 1, 2, 3, 4$. It's trivial to calculate $T(1)$:

$$T(1) \equiv 1^{25} \equiv 1 \mod 125.$$

Similarly, since $4 \equiv -1 \bmod 5$ we can easily calculate $T(4)$:

$$T(4) \equiv (-1)^{25} \equiv -1 \mod 125.$$

We showed earlier that $T(37) \equiv 57 \bmod 125$. Since $2 \equiv 37 \bmod 5$, we must also have

$$T(2) \equiv 57 \mod 125.$$

Finally, since T is a homomorphism we have

$$T(3) \equiv T(-2) \equiv T(-1)T(2) \equiv -57 \equiv 68 \mod 125.$$

Therefore, the Teichmüller lifts modulo 125 are:

$x \bmod 5$	$T(x) \bmod 125$
1	1
2	57
3	68
4	124

Proof of Proposition 4.4. Let $x \in (\mathbb{Z}/p^n)^\times$. By Euler's Theorem, we have $x^{(p-1)p^{n-1}} \equiv 1 \bmod p^n$. We can rewrite this as

$$T(x)^{p-1} \equiv 1 \mod p^n,$$

which proves (2). Hence, $T(x)^p \equiv T(x) \bmod p^n$, so the sequence x^{p^r} stabilizes at $T(x)$, which proves (1).

To prove (3), suppose $x \equiv y \bmod p$. By Lemma 4.6, we know $x^p \equiv y^p \bmod p^2$. Using the lemma again, we get $x^{p^2} \equiv y^{p^2} \bmod p^3$ etc. After using the lemma $n-1$ times, we end up with $T(x) \equiv T(y) \bmod p^n$. To prove the congruence $T(x) \equiv x \bmod p$, we use (1) with $n = 1$.

To show that T is a homomorphism, note that

$$T(xy) \equiv (xy)^{p^{n-1}} \equiv x^{p^{n-1}} y^{p^{n-1}} \equiv T(x)T(y) \mod p^n.$$

To show that $T : \mathbb{F}_p^\times \to (\mathbb{Z}/p^n)^\times$ is injective, suppose that $T(x) \equiv T(y) \bmod p^n$. By (3), we know that $T(x) \equiv x \bmod p$ and $T(y) \equiv y \bmod p$. Hence, $x \equiv y \bmod p$. □

Corollary 4.2. *Let p be an odd prime number. Every element of $(\mathbb{Z}/p^n)^\times$ can be written uniquely in the form $T(x) \cdot \exp(py)$, where $x \in \mathbb{F}_p^\times$ and $py \in p\mathbb{Z}/p^n$. In other words, we have an isomorphism of groups:*

$$(\mathbb{Z}/p^n)^\times \cong \mathbb{F}_p^\times \times p\mathbb{Z}/p^n.$$

(For clarity, we emphasize that it is the number py which is unique modulo p^n in this corollary. This means that y is uniquely defined modulo p^{n-1}.)

Proof (*Existence*). Take any $a \in (\mathbb{Z}/p^n)^\times$ and define $x \in \mathbb{F}_p^\times$ by the condition $x \equiv a \bmod p$. This implies $T(x) \equiv a \bmod p$, and therefore $T(x)^{-1}a \equiv 1 \bmod p$. Hence, $T(x)^{-1}a$ is in the subgroup $1 + p\mathbb{Z}/p^n$, so we can set $py \equiv \log(T(x)^{-1}a) \bmod p^n$. Obviously, $a \equiv T(x)\exp(py) \bmod p^n$.

(*Uniqueness*). Suppose $T(x) \cdot \exp(py) \equiv T(x') \cdot \exp(py') \bmod p^n$. Since the image of exp consists of numbers congruent to 1 modulo p, we have

$$T(x) \equiv T(x') \mod p.$$

By part 3 of Proposition 4.4, we have $T(x) \equiv x \bmod p$ and $T(x') \equiv x' \bmod p$. It follows that $x \equiv x' \bmod p$ and therefore $T(x) \equiv T(x') \bmod p^n$. Cancelling $T(x)$ from our equation, we have $\exp(py) \equiv \exp(py') \bmod p^n$. Taking logarithms, we get $py \equiv py' \bmod p^n$. □

As long as the prime number p is relatively small, Corollary 4.2 allows us to do calculations in the group $(\mathbb{Z}/p^n)^\times$ with the same ease that we did calculations in $1+p\mathbb{Z}/p^n$ in the previous section. As an example, we'll write the number 22 as $T(x)\exp(py)$ modulo 125. Note that $22 \equiv 2 \bmod 5$, so we have $x = 2$. This implies

$$\exp(py) \equiv 22 \times T(2)^{-1} \mod 125.$$

Since $2^{-1} \equiv 3 \bmod 5$, we have $T(2)^{-1} \equiv T(3) \bmod 125$. From the table above, we have $T(3) \equiv 68 \bmod 125$. Substituting this, we get:

$$\exp(py) \equiv 22 \times 68 \equiv -4 \mod 125.$$

Therefore, $py \equiv \log(-4) \bmod 125$. The power series expansion of $\log(1-5)$ is as follows:

$$py \equiv -5 - \frac{5^2}{2} - \frac{5^3}{3} - \cdots \mod 125.$$

Only the first two terms of this series are non-zero modulo 5^3. Since $2^{-1} \equiv 3 \bmod 5$, we have $\frac{5^2}{2} \equiv 75 \bmod 125$. This implies

$$py \equiv -5 - 75 \equiv 45 \mod 125.$$

We've shown that

$$22 \equiv T(2) \cdot \exp(45) \mod 125. \tag{4.17}$$

Calculating powers. We can use the decomposition in Equation (4.17) to calculate powers of 22 modulo 125 very quickly. For example, to calculate 22^{37} we begin with

$$22^{37} \equiv T(2^{37}) \cdot \exp(37 \times 45) \mod 125.$$

By Fermat's Little Theorem, we know that $2^4 \equiv 1 \bmod 5$. This implies $2^{37} \equiv 2 \bmod 5$, and therefore

$$T(2^{37}) \equiv T(2) \equiv 57 \mod 125.$$

Also, since $37 \cdot 45 \equiv 40 \bmod 125$, we have

$$\exp(37 \cdot 45) \equiv \exp(40) \equiv 1 + 40 + \frac{40^2}{2} \mod 125.$$

It follows that 22^{37} is equal to

$$T(2) \cdot \exp(40) \equiv 57 \cdot 841 \equiv 62 \mod 125.$$

Calculating the order of an element. As another easy application of Equation (4.17), we'll calculate the order of 22 as an element of the group $(\mathbb{Z}/125)^\times$. This is the smallest integer n such that $22^n \equiv 1 \bmod 125$. We can rewrite the equation as follows:

$$T(2^n) \cdot \exp(45n) \equiv T(1)\exp(0) \mod 125.$$

By the uniqueness part of Corollary 4.2, this is equivalent to two simultaneous equations:

$$2^n \equiv 1 \bmod 5 \quad \text{and} \quad 45n \equiv 0 \bmod 125.$$

We can easily check that 2 has order 4 in $\mathbb{F}_5^\times$, so the first equation is equivalent to $n \equiv 0 \bmod 4$. Solving the second equation, we get $n \equiv 0 \bmod 25$. Since 4 and 25 are coprime, the Chinese Remainder Theorem implies that $n \equiv 0 \bmod 100$. This shows that 22 has order 100 in $(\mathbb{Z}/125)^\times$. In particular, we see that $(\mathbb{Z}/125)^\times$ is a cyclic group, and is generated by 22.

Solving congruences with powers. As another example, we'll solve the congruence

$$x^{87} \equiv 22 \mod 125.$$

We could solve this congruence by the method of Section 3.2, but the solution would be $x \equiv 22^{87^{-1} \bmod \varphi(125)}$. Since $87^{-1} \equiv 23 \bmod \varphi(125)$, it's clear that solving the congruence by this method would take a lot of paper. However, if we instead write x in the form $T(a) \cdot \exp(5b)$, then the congruence can be rewritten:

$$T(a^{87}) \cdot \exp(87 \cdot 5 \cdot b) \equiv T(2) \cdot \exp(45) \mod 125.$$

By the uniqueness part of Corollary 4.2, this is equivalent to the two simultaneous congruences:

$$a^{87} \equiv 2 \bmod 5 \quad \text{and} \quad 87 \cdot 5 \cdot b \equiv 45 \bmod 125.$$

Using Fermat's Little Theorem, we can reduce the first of these congruences to $a \equiv 2^{87^{-1} \bmod 4}$. Since $87 \equiv 3 \bmod 4$, we have $87^{-1} \equiv 3 \bmod 4$, and therefore

$$a \equiv 2^3 \equiv 3 \mod 5.$$

The second of the simultaneous congruences is linear, so we can solve it very quickly. Since $87 \equiv 12 \bmod 25$, we have $87^{-1} \equiv -2 \bmod 25$, and therefore

$$5b \equiv 45 \cdot (-2) \equiv 35 \mod 125.$$

We can now recover the solution x from a and b as follows:

$$x \equiv T(3) \exp(35) \equiv 68 \cdot \left(1 + 35 + \frac{5^2 \cdot 7^2}{2}\right) \mod 125.$$

Since $7^2 \equiv 4 \bmod 5$, we have $\frac{5^2 \cdot 7^2}{2} \equiv 50 \bmod 125$. Therefore,

$$x \equiv 68 \cdot 86 \equiv 98 \mod 125.$$

The multiplicative group $(\mathbb{Z}/n)^\times$. We've seen that the decomposition of $(\mathbb{Z}/p^n)^\times$ as a direct product of the cyclic groups $\mathbb{F}_p^\times$ and $p\mathbb{Z}/p^n$ allows us to perform calculations in the group $(\mathbb{Z}/p^n)^\times$ very quickly. To complete this discussion, we consider also the case $p = 2$. The theorems of this section and the previous section are not valid in this case. It's not just because the proofs are more difficult; it's actually because the results are false. For example, $\exp(2x)$ does not converge 2-adically for all $x \in \mathbb{Z}_{(2)}$. Furthermore, since $\mathbb{F}_2^\times = \{1\}$, the only "Teichmüller lift" we could consider would be $T(1) = 1$. However, it is true that $\exp(4x)$ converges 2-adically for all $x \in \mathbb{Z}_{(2)}$, as does $\log(1+4x)$. Furthermore, these functions are inverses and give an isomorphism between the additive and multiplicative groups:

$$1 + 4\mathbb{Z}/2^n \cong 4\mathbb{Z}/2^n.$$

Furthermore, for $n \geq 2$, every element of $(\mathbb{Z}/2^n)^\times$ can be written uniquely in the form $T \cdot \exp(4y)$, where $T = \pm 1$ (see Exercise 4.34). In particular, we have a decomposition:

$$(\mathbb{Z}/2^n)^\times \cong \{1, -1\} \times 4\mathbb{Z}/2^n.$$

In the case of a general multiplicative group $(\mathbb{Z}/n)^\times$, where n is not necessarily a power of a prime, there is a similar decomposition. Recall first that as a consequence of the Chinese Remainder Theorem, we proved in Theorem 3.4 that when n and m are coprime we have an isomorphism

$$(\mathbb{Z}/nm)^\times \cong (\mathbb{Z}/n)^\times \times (\mathbb{Z}/m)^\times.$$

This isomorphism takes $x \in (\mathbb{Z}/nm)^\times$ to the pair $(x \bmod n, x \bmod m)$, and its inverse takes the pair (x, y) to $hny + kmx$, where the integers h and k are chosen to satisfy the equation $hn + km = 1$. If we factorize n as $n = p_1^{a_1} \cdots p_r^{a_r}$, then we'll get a decomposition of $(\mathbb{Z}/n)^\times$:

$$(\mathbb{Z}/n)^\times \cong (\mathbb{Z}/p_1^{a_1})^\times \times \cdots \times (\mathbb{Z}/p_r^{a_r})^\times.$$

This allows us to decompose each multiplicative group as a product of cyclic groups, which are either $\mathbb{F}_p^\times$ or additive. For example, since $325 = 3^4 \cdot 5$, we have

$$(\mathbb{Z}/325)^\times \cong \mathbb{F}_3^\times \times 3\mathbb{Z}/81 \times \mathbb{F}_5^\times.$$

In particular, as long as the prime factors of n are relatively small, we can do arithmetic in $(\mathbb{Z}/n)^\times$ very quickly.

Exercise 4.27. Find all Teichmüller lifts in $(\mathbb{Z}/3^{100})^\times$.

Exercise 4.28.

(1) Find all the Teichmüller lifts modulo 5^4.
(2) Find x, y such that $13 \equiv T(x)\exp(5y) \bmod 5^4$.
(3) Hence, calculate $13^{234} \bmod 5^4$.
(4) Solve the congruence $x^{47} \equiv 13 \bmod 5^4$.

Exercise 4.29.

(1) Find the Teichmüller lift of 2 in $\mathbb{Z}/11^3$.
(2) Hence, calculate all the Teichmüller lifts modulo 11^3.
(3) Write each of the following numbers in the form $T(x)\exp(11y)$ modulo 11^3:

$$6, \quad 17, \quad 35, \quad 47.$$

Exercise 4.30. Find all solutions in $\mathbb{Z}/7^{30}$ to each of the following congruences, expressing each solution in the form $T(a)\exp(7b)$:

$$x^4 \equiv \exp(112) \mod 7^{30}, \qquad x^7 \equiv T(3)\exp(98) \mod 7^{30}.$$

Exercise 4.31. Let p be a prime number, such that $p \equiv 3 \bmod 4$. Show that there are exactly two solutions in $\mathbb{Z}/p^n$ to the congruence $x^8 \equiv 1 \bmod p^n$.

Exercise 4.32.

(1) Let p be an odd prime and let g be a primitive root modulo p. Show that $T(g)\exp(p)$ is a generator of $(\mathbb{Z}/p^n)^\times$, and in particular $(\mathbb{Z}/p^n)^\times$ is cyclic.
(2) Let G and H be finite groups. Show that an element $(x, y) \in G \times H$ is a generator of $G \times H$ if and only if x is a generator of G, y is a generator of H, and $\text{hcf}(|G|, |H|) = 1$. Hence, show that $G \times H$ is cyclic if and only if G and H are both cyclic and the orders of G and H are coprime.
(3) For which positive integers n is the group $(\mathbb{Z}/n)^\times$ cyclic?

Exercise 4.33. Let p be an odd prime and let $f(X) = X^{p-1} - 1$.

(1) For each integer a_0 which is not a multiple of p, show that a_0 satisfies the conditions of Hensel's Lemma.
(2) Show that $T(a_0) \equiv a_n \bmod p^{2^n}$, where T is the Teichmüller lift and a_n is the sequence defined in Hensel's Lemma.

Exercise 4.34.

(1) Assume $n \geq 3$. Show that every element of $(\mathbb{Z}/2^n)^\times$ can be written uniquely as $T \cdot \exp(4x)$ for some $4x \in 4\mathbb{Z}/2^n$ and $T = \pm 1$.
(2) Show that for $n \geq 3$, the congruence $x^2 - 1 \equiv 0 \bmod 2^n$ has exactly four solutions modulo 2^n. Find these four solutions. Hence, write down two distinct factorizations of the polynomial $x^2 - 1$ over $\mathbb{Z}/2^n$.

Exercise 4.35. Let p be a prime number, let $\zeta_p = e^{\frac{2\pi i}{p}}$ and let $f(x)$ be a polynomial with coefficients in $\mathbb{Z}$, and with a root in $\mathbb{Z}[\zeta_p]$. Show that for every prime $q \equiv 1 \bmod p$ and every $n \geq 0$, the congruence $f(x) \equiv 0 \bmod q^n$ has a solution.

4.6 THE RING OF p-ADIC INTEGERS

In this section, we'll look at the results of this chapter in a new way. We've defined several congruency classes such as $\exp(px) \bmod p^n$, $T(a) \bmod p^n$ and $\log(1+px) \bmod p^n$. It would be convenient to be able to write down just $\exp(px)$, $T(a)$, etc., without needing to write "modulo p^n" each time. The problem is that there is no integer (or even an element of the local ring) which is congruent to $T(a)$ modulo p^n for all n, and so $T(a)$ does not have a meaning in $\mathbb{Z}_{(p)}$. Because of this, we shall introduce a bigger ring, called the ring $\mathbb{Z}_p$ of p-adic integers. The expressions $T(a)$, $\exp(px)$ etc. all make sense in the ring of p-adic integers. Roughly speaking, a p-adic integer is a p-adically convergent series, in the sense defined in Section 4.3; two series are equal as p-adic integers if they are congruent modulo all powers of p.

Construction of the real numbers. It's often useful in mathematics to use numbers such as $\sqrt{2}$ or π in calculations. However, $\sqrt{2}$ and π are not in the field $\mathbb{Q}$. We might get away with using approximations to π, such as $\frac{22}{7}$ or 3.14, but it's easier and more accurate to simply write down π in a formula. To be able to do this, we need to use the field of real numbers, rather than the rational numbers. We'll discuss the construction of the real numbers briefly, because it is similar to the construction of the p-adic integers. For the moment we should pretend that we have never heard of the real numbers, and only know the field of rational numbers.

A sequence x_n of rational numbers is called a *Cauchy sequence* if it satisfies the condition[b]

$$\forall \epsilon > 0 \ \exists N \text{ such that } n, m > N \implies |x_n - x_m| < \epsilon.$$

We would like the Cauchy sequences to have limits. However, many of them do not have rational limits, and so from our current point of view they do not have limits at all. We'll say that two Cauchy sequences x_n and y_n are

---

[b]Technically speaking, we should assume that ϵ is a positive rational number, since we have not yet constructed the real numbers.

equivalent if

$$\forall \epsilon > 0 \ \exists N \text{ such that } n > N \implies |x_n - y_n| < \epsilon.$$

Intuitively, this means that x_n and y_n should have the same limit (if only such a limit existed). One can easily show that equivalence of Cauchy sequences is an equivalence relation, and we define the real numbers $\mathbb{R}$ to be the set of equivalence classes of Cauchy sequences of rational numbers.

So far, we've just defined the real numbers as a set. However, we can also add and multiply real numbers. If $\underline{x}$ and $\underline{y}$ are real numbers represented by the Cauchy sequences x_n and y_n, then we may define $\underline{x} + \underline{y}$ and $\underline{x} \cdot \underline{y}$ to be the real numbers represented by the Cauchy sequences $(x_n + y_n)$ and $(x_n \cdot y_n)$, respectively. Of course, to make sense of this definition we need to prove the following statements, both of which are quite easy:

(1) If x_n and y_n are Cauchy sequences, then $x_n + y_n$ and $x_n \cdot y_n$ are also Cauchy sequences.
(2) If x_n is equivalent to x'_n and y_n is equivalent to y'_n, then $x_n + y_n$ is equivalent to $x'_n + y'_n$ and $x_n \cdot y_n$ is equivalent to $x'_n \cdot y'_n$.

Once this is done, it is not hard to prove that the real numbers form a field. We can think of the rational numbers as being contained in the real numbers, by identifying a rational number x with the equivalence class of the constant sequence $(x, x, x, \ldots)$.

***p*-Adic Cauchy sequences.** For a rational number x, we define the *p-adic absolute value* $|x|_p$ as follows:

$$|x|_p = \begin{cases} p^{-v_p(x)} & \text{if } x \neq 0, \\ 0 & \text{if } x = 0. \end{cases}$$

Note that $|x|_p$ is close to zero if x is in $\mathbb{Z}_{(p)}$ and x is congruent to zero modulo a large power of p. Hence, $|x|_p$ is just another notation for dealing with congruences. However, its similarity to the notation for the absolute value of a real number makes certain statements in number theory appear very much like familiar statements in analysis. For example, we know from number theory that if $x \equiv 0 \bmod p^r$ and $y \equiv 0 \bmod p^r$, then $x + y \equiv 0 \bmod p^r$. In the new notation, this becomes

$$|x + y|_p \leq \max\{|x|_p, |y|_p\}. \tag{4.18}$$

Equation (4.18) is called the *ultrametric inequality*; it is a strong version of the "triangle inequality". Furthermore, by uniqueness of factorization in $\mathbb{Z}$, we have $v_p(x \cdot y) = v_p(x) + v_p(y)$. This implies

$$|x \cdot y|_p = |x|_p \cdot |y|_p.$$

Imitating the construction of the real numbers, we'll call a sequence $x_n \in \mathbb{Z}_{(p)}$ a *p-adic Cauchy sequence* if it satisfies the following condition:

$$\forall \epsilon > 0 \ \exists N \text{ such that } n, m \geq N \implies |x_n - x_m|_p \leq \epsilon. \tag{4.19}$$

In the language of congruences, this means that for any $r \in \mathbb{N}$ there exists an integer N such that

$$x_N \equiv x_{N+1} \equiv x_{N+2} \equiv \cdots \mod p^r.$$

We have seen examples of p-adic Cauchy sequences throughout this chapter. For example, the sequence x^{p^n} defining the Teichmüller lift is a p-adic Cauchy sequence. Indeed, we have shown in Proposition 4.4 that

$$x^{p^{r-1}} \equiv x^{p^r} \equiv x^{p^{r+1}} \equiv \cdots \mod p^r.$$

Next, suppose that a_0 satisfies the conditions of Hensel's Lemma for a polynomial f. This means that we have $f(a_0) \equiv 0 \bmod p^{2c+1}$, where $c = v_p(f'(a_0))$. We defined a sequence a_n by the Newton–Raphson formula:

$$a_{n+1} = a_n - \delta, \quad \delta = \frac{f(a_n)}{f'(a_n)}.$$

It turns out that the sequence a_n is also a p-adic Cauchy sequence. To see why this is true, recall from Equation (4.3) that $v_p(\delta) \geq c+2^n$. This implies the congruence

$$a_{n+1} \equiv a_n \mod p^{c+2^n}.$$

If we choose N large enough so that $c + 2^N \geq r$, then we have

$$a_N \equiv a_{N+1} \equiv a_{N+2} \equiv \cdots \mod p^r.$$

As another example, suppose we have a series $\sum x_n$ in $\mathbb{Z}_{(p)}$, which converges p-adically in the sense of Section 4.3. Let S_n be the nth partial sum of the series, i.e.

$$S_n = x_1 + x_2 + \cdots + x_n.$$

Since $\sum x_n$ converges p-adically, there is an N such that all the terms $x_N, x_{N+1}, \cdots$ are multiples of p^r. This implies

$$S_N \equiv S_{N+1} \equiv S_{N+2} \equiv \cdots \mod p^r.$$

In other words, the sequence of partial sums is a p-adic Cauchy sequence.

If $\underline{x} = (x_n)$ is a p-adic Cauchy sequence, then for any r, the sequence of congruency classes ($x_n \bmod p^r$) is eventually constant in $\mathbb{Z}/p^r$. We call this constant the reduction of $\underline{x}$ modulo p^r. If $\underline{x} = (x_n)$ and $\underline{y} = (y_n)$ converge to the same constant value modulo p^r, then we say $\underline{x} \equiv \underline{y} \bmod p^r$. This means that we have $x_n \equiv y_n \bmod p^r$ for n sufficiently large.

Construction of the p-adic integers. To construct the ring of p-adic integers $\mathbb{Z}_p$, we shall add to $\mathbb{Z}_{(p)}$ the limits of all the p-adic Cauchy sequences. More formally, we call two Cauchy sequences $\underline{x} = (x_n)$ and $\underline{y} = (y_n)$ equivalent if they satisfy the following analytic condition:

$$\forall \epsilon > 0 \ \exists N \text{ such that } n \geq N \implies |x_n - y_n|_p \leq \epsilon.$$

In terms of congruences, the Cauchy sequences x_n and y_n are equivalent if

$$\forall r \in \mathbb{N} \quad \underline{x} \equiv \underline{y} \bmod p^r.$$

We define the p-adic integers $\mathbb{Z}_p$ to be the set of equivalence classes of p-adic Cauchy sequences in $\mathbb{Z}_{(p)}$.

For example, if a series $\sum x_n$ converges p-adically, then its sequence of partial sums is a p-adic Cauchy sequence. If we just write $\sum x_n$, without writing "modulo p^r" then we shall mean the p-adic integer represented by the sequence of partial sums. Note that the series $\sum x_n$ and $\sum y_n$ are equal in $\mathbb{Z}_p$ if and only if their sequences of partial sums are equivalent. This is the same as saying that $\sum x_n \equiv \sum y_n \bmod p^r$ for all r. This notation gives us the following useful way of representing p-adic integers.

Proposition 4.5. *Every p-adic integer can be written uniquely in the form $\sum_{r=0}^{\infty} a_r p^r$, with coefficients $a_r \in \{0, 1, \ldots, p-1\}$.*

Intuitively, Proposition 4.5 says that we can write a p-adic integer in base p using the digits $0, 1, \ldots, p-1$, but this expansion may be infinite to the left (as opposed to having infinitely many digits to the right, as real numbers have). It also says that such an expansion is unique. We therefore avoid the annoying phenomenon in $\mathbb{R}$, where one real number can have two different decimal expansions:

$$1.0000000 \cdots = 0.99999999 \cdots.$$

Proof of Proposition 4.5 (*Existence*). Let $\underline{x}$ be a p-adic integer represented by a Cauchy sequence (x_n). For sufficiently large N, the numbers $x_N, x_{N+1}, x_{N+2}, \ldots$ are all congruent modulo p. If we choose a_0 to be in the congruency class of x_N, then the numbers $x_N - a_0, x_{N+1} - a_0, \ldots$ are all multiples of p.

By increasing N is necessary, we may assume that the numbers $x_N - a_0, x_{N+1} - a_0, \ldots$ are all congruent modulo p^2. Since these numbers are all multiples of p, they are all congruent to pa_1 for some choice of a_1. Hence, the numbers

$$x_N - a_0 - a_1 p, \quad x_{N+1} - a_0 - a_1 p, \quad x_{N+2} - a_0 - a_1 p, \ldots$$

are all multiples of p^2. Note that this means $\underline{x} \equiv a_0 + a_1 p \bmod p^2$.

Continuing in this way, we form a sequence a_n, such that for any r we have:

$$\underline{x} \equiv a_0 + a_1 p + \cdots + a_{r-1} p^{r-1} \mod p^r.$$

The series $\sum a_n p^n$ converges p-adically, since $v_p(a_n p^n) \geq n$. Therefore, $\sum a_n p^n$ represents a p-adic integer. The congruence above can be rewritten as

$$\underline{x} \equiv \sum_{n=0}^{\infty} a_n p^n \mod p^r.$$

Since this congruence holds for all r, we have $\underline{x} = \sum a_n p^n$.

(*Uniqueness*). Suppose $\sum a_n p^n = \sum b_n p^n$ in $\mathbb{Z}_p$, with coefficients a_n and b_n in $\{0, 1, \ldots, p-1\}$. Reducing this equation modulo p we get $a_0 \equiv b_0 \bmod p$, and this implies $a_0 = b_0$. Reducing modulo p^2 instead of modulo p, we get

$$a_0 + a_1 p \equiv b_0 + b_1 p \mod p^2.$$

We've already seen that $a_0 = b_0$, so we have $pa_1 \equiv pb_1 \bmod p^2$. Cancelling the p we get $a_1 \equiv b_1 \bmod p$, and this implies $a_1 = b_1$. Continuing in this way, we see that $a_n = b_n$ for all n. □

To make $\mathbb{Z}_p$ into a ring, we need to define addition and multiplication operations. We'd like to define these operations in terms of Cauchy sequences as follows:

$$(x_n) + (y_n) = (x_n + y_n), \quad (x_n) \cdot (y_n) = (x_n \cdot y_n).$$

We need to check several things before we can use this as a definition. First, we need to check that if (x_n) and (y_n) are p-adic Cauchy sequences, then the sequences $(x_n + y_n)$ and $(x_n \cdot y_n)$ are also p-adic Cauchy sequences. Secondly, we need to show that if (x_n) is equivalent to (x'_n) and (y_n) is equivalent to (y'_n), then $(x_n + y_n)$ is equivalent to $(x'_n + y'_n)$ and $(x_n \cdot y_n)$ is equivalent to $(x'_n \cdot y'_n)$. All of this is left as Exercise 4.37. Once this is done, we have well defined operations on the set $\mathbb{Z}_p$. We also need to verify the

Ring Axioms. For example, we need to show that $(x_n) + (y_n)$ is equivalent to $(y_n) + (x_n)$. However, these axioms are all obvious from the definitions.

We may think of the local ring $\mathbb{Z}_{(p)}$ as a subring of $\mathbb{Z}_p$ by identifying x with the constant sequence $(x, x, x, x, x, \ldots)$. However, it turns out that there are many more p-adic integers than there are elements in the local ring $\mathbb{Z}_{(p)}$. For example, consider the following 5-adic integer:

$$\underline{x} = (1+5)^{\frac{1}{2}} = 1 + \frac{1}{2} \cdot 5 + \frac{(\frac{1}{2})(-\frac{1}{2})}{2} \cdot 5^2 + \cdots .$$

By this, we mean that $\underline{x}$ is the equivalence class of the sequence of partial sums of the series above. We can show by the Power Series Trick that $\underline{x}^2 \equiv 6 \bmod 5^r$ for all r, and therefore $\underline{x}^2 = 6$ in $\mathbb{Z}_5$. However, the local ring $\mathbb{Z}_{(5)}$ has no square roots of 6, since its elements are rational numbers. This shows that $\underline{x}$ is in $\mathbb{Z}_5$ but not $\mathbb{Z}_{(5)}$. The following result gives many more examples of p-adic integers which are not in $\mathbb{Z}_{(p)}$.

Proposition 4.6. *Let a_0 and f satisfy the conditions of Hensel's Lemma. Then there is an $\underline{a} \in \mathbb{Z}_p$ such that $f(\underline{a}) = 0$.*

For example, we can show that $\mathbb{Z}_5$ contains a root of the polynomial $X^4 + 5X^3 + 15X + 4$. This is because $a_0 = 1$ satisfies the conditions of Hensel's Lemma.

Proof. We've seen above that the sequence a_n in Hensel's Lemma is a p-adic Cauchy sequence. Let $\underline{a}$ be the equivalence class of this sequence. Fix any $r \in \mathbb{N}$. For n sufficiently large, we have $\underline{a} \equiv a_n \bmod p^r$. By increasing n if necessary, we may assume that $2c + 2^n \geq r$, and therefore $f(a_n) \equiv 0 \bmod p^r$. This implies $f(\underline{a}) \equiv 0 \bmod p^r$. As this congruence holds for all r, it follows that $f(\underline{a}) = 0$ in $\mathbb{Z}_p$. □

Next consider Teichmüller lifts. For an odd prime p and an element $x \in \mathbb{Z}_{(p)}$ which is not a multiple of p, we saw that the sequence x^{p^n} is a p-adic Cauchy sequence, so we may define a p-adic integer $\underline{T}(x)$ to be the equivalence class of this sequence. The limiting congruency classes $T(x) \bmod p^r$, which we defined in Section 4.5, are the reductions of $\underline{T}(x)$ modulo p^r.

Proposition 4.7. *The p-adic integer $\underline{T}(x)$ depends only on x modulo p, and the map $\underline{T} : \mathbb{F}_p^\times \to \mathbb{Z}_p^\times$ is an injective group homomorphism.*

Proof. Suppose $x \equiv y \bmod p$. Fix any $r \in \mathbb{N}$. By Proposition 4.4, it follows that $x^{p^n} \equiv y^{p^n} \bmod p^r$ whenever $n \geq r - 1$. This shows that the sequences

x^{p^n} and y^{p^n} are equivalent, so $\underline{T}(x) = \underline{T}(y)$. Since $\underline{T}(x) \equiv x \bmod p$, it follows that $\underline{T}$ is injective on $\mathbb{F}_p^\times$. Clearly $(xy)^{p^n} = x^{p^n} \cdot y^{p^n}$, and therefore $\underline{T}(xy) = \underline{T}(x)\underline{T}(y)$. □

The p-adic absolute value on $\mathbb{Z}_p$. Suppose $\underline{x} = (x_n)$ is a non-zero p-adic integer. This means that there is some r such that $\underline{x} \not\equiv 0 \bmod p^r$. In other words, the limiting congruency class in the sequence $x_n \bmod p^r$ is a non-zero element in $\mathbb{Z}/p^r$. We may therefore define the p-adic valuation of $\underline{x}$ as follows:

$$v_p(\underline{x}) = \begin{cases} \max\{r : \underline{x} \equiv 0 \bmod p^r\} & \text{if } \underline{x} \neq 0, \\ \infty & \text{if } \underline{x} = 0. \end{cases}$$

This coincides with our definition of v_p on the local ring $\mathbb{Z}_{(p)}$. In fact, for N sufficiently large, we have

$$v_p(\underline{x}) = v_p(x_N) = v_p(x_{N+1}) = v_p(x_{N+2}) = v_p(x_{N+3}) = \cdots .$$

If we write down the base p expansion of $\underline{x}$ as in Proposition 4.5, then $v_p(\underline{x})$ is the number of zeros at the end of the expansion. We may also define the p-adic absolute value on $\mathbb{Z}_p$:

$$|\underline{x}|_p = \begin{cases} p^{-v_p(\underline{x})} & \text{if } \underline{x} \neq 0, \\ 0 & \text{if } \underline{x} = 0. \end{cases}$$

Proposition 4.8. *For $\underline{x}$ and $\underline{y}$ in $\mathbb{Z}_p$, we have $|\underline{x} \cdot \underline{y}|_p = |\underline{x}|_p \cdot |\underline{y}|_p$ and $|\underline{x} + \underline{y}|_p \leq \max\{|\underline{x}|_p, |\underline{y}|_p\}$.*

Proof. We already know that these equations are true on $\mathbb{Z}_{(p)}$. To see that they remain true on $\mathbb{Z}_p$, note that $v_p(\underline{x}) = v_p(x_n)$ for n large enough. □

Proposition 4.9. $\mathbb{Z}_p^\times = \{\underline{x} \in \mathbb{Z}_p : |\underline{x}|_p = 1\}$.

Proof. Suppose $\underline{x}$ has an inverse $\underline{x}^{-1} \in \mathbb{Z}_p$. Since $\underline{x} \cdot \underline{x}^{-1} = 1$, we have $v_p(\underline{x}) + v_p(\underline{x}^{-1}) = v_p(1) = 0$, and therefore $v_p(\underline{x}) = 0$.

Conversely, suppose $v_p(\underline{x}) = 0$. For n sufficiently large, we have $v_p(x_n) = 0$, where x_n is a p-adic Cauchy sequence representing $\underline{x}$. By changing finitely many terms, we may replace x_n by an equivalent sequence, all of whose terms have valuation 0. But then $x_n^{-1} \in \mathbb{Z}_{(p)}$ for all n. Since x_n is a p-adic Cauchy sequence, we have for N sufficiently large

$$x_N \equiv x_{N+1} \equiv x_{N+2} \equiv \cdots \mod p^r.$$

This implies

$$x_N^{-1} \equiv x_{N+1}^{-1} \equiv x_{N+2}^{-1} \equiv \cdots \mod p^r,$$

and therefore x_n^{-1} is also a p-adic Cauchy sequence. The equivalence class of (x_n^{-1}) is the inverse of $\underline{x}$. □

As we have defined the p-adic absolute value on the whole of $\mathbb{Z}_p$, we can now say what it means for a sequence in $\mathbb{Z}_p$ to converge. If $\underline{x}_n$ is a sequence of elements of $\mathbb{Z}_p$ and $\underline{y}$ is an element of $\mathbb{Z}_p$, then we say that $\underline{y}$ is the limit of the sequence $\underline{x}_n$ if

$$\forall \epsilon > 0 \ \exists N \text{ such that } n \geq N \implies |\underline{x}_n - \underline{y}|_p < \epsilon.$$

The p-adic integers $\underline{x}_n$ and $\underline{y}$ represent congruency classes modulo p^r for all r. The convergence condition can be expressed in the language of congruences as follows:

$$\forall r \in \mathbb{N} \ \exists N \text{ such that } \underline{y} \equiv \underline{x}_N \equiv \underline{x}_{N+1} \equiv \underline{x}_{N+2} \equiv \cdots \bmod p^r.$$

The next proposition shows that $\mathbb{Z}_p$ is exactly the set of limits of sequences in $\mathbb{Z}_{(p)}$.

Proposition 4.10. *Let $\underline{x}$ be a p-adic integer represented by the p-adic Cauchy sequence x_n in $\mathbb{Z}_{(p)}$. Then x_n converges to $\underline{x}$ in $\mathbb{Z}_p$.*

Proof. Fix any $r \in \mathbb{N}$. Since x_n is a Cauchy sequence, there is an N such that $x_N \equiv x_{N+1} \equiv \cdots \bmod p^r$. This limiting congruency class is by definition the congruency class $\underline{x} \bmod p^r$. We therefore have $x_n \equiv \underline{x} \bmod p^r$ for all $n \geq N$. □

It's quite easy to test whether a series converges in $\mathbb{Z}_p$; the following result generalizes Lemma 4.3.

Theorem 4.3. *Let $\sum \underline{x}_n$ be a series of elements of $\mathbb{Z}_p$. If $|\underline{x}_n|_p \to 0$, then $\sum \underline{x}_n$ converges in $\mathbb{Z}_p$.*

Proof. We'll write $\underline{S}_n = \underline{x}_1 + \cdots + \underline{x}_n$ for the n-th partial sum. We must show that the sequence $\underline{S}_n$ converges in $\mathbb{Z}_p$. For any n we can choose a $y_n \in \mathbb{Z}_{(p)}$, such that $y_n \equiv \underline{S}_n \bmod p^n$. If we fix an r, then for large N the numbers $\underline{x}_N, \underline{x}_{N+1}, \ldots$ are all congruent to 0 modulo p^r, and therefore $\underline{S}_{N-1} \equiv \underline{S}_N \equiv \cdots \bmod p^r$. By increasing N if necessary, we may assume that $N > r$. This implies $y_{N-1} \equiv y_N \equiv \cdots \bmod p^r$, so y_n is a p-adic Cauchy sequence. By Proposition 4.10, we know that y_n converges to $\underline{y}$. Since $|y_n - \underline{S}_n|_p \leq p^{-n}$, we can show by the ultrametric inequality that $\underline{S}_n$ also converges to $\underline{y}$. □

For odd p, the theorem shows that the power series $\exp(p\underline{x})$ and $\log(1+p\underline{x})$ both converge for all $\underline{x} \in \mathbb{Z}_p$. Using the power series trick, one can show that the functions exp and log give inverse isomorphisms between the additive group $p\mathbb{Z}_p$ and the multiplicative group $1+p\mathbb{Z}_p$. Furthermore, there is an isomorphism

$$\mathbb{Z}_p^\times \cong \mathbb{F}_p^\times \times p\mathbb{Z}_p.$$

The map from $\mathbb{F}_p^\times \times p\mathbb{Z}_p$ to $\mathbb{Z}_p^\times$ is $(x, p\underline{y}) \mapsto \underline{T}(x)\exp(p\underline{y})$.

The p-adic numbers $\mathbb{Q}_p$. We may construct a completion $\mathbb{Q}_p$ of the field $\mathbb{Q}$ by exactly the same process. The elements of $\mathbb{Q}_p$ are the equivalence classes of p-adic Cauchy sequences of rational numbers. Elements of $\mathbb{Q}_p$ are called p-adic numbers. We define addition and multiplication of these elements by adding and multiplying the sequences. The ring $\mathbb{Z}_p$ is contained in $\mathbb{Q}_p$, but $\mathbb{Q}_p$ is bigger; for example, it contains $\mathbb{Q}$, so in particular it contains the number $\frac{1}{p}$ which is not in $\mathbb{Z}_p$. In general, if $x_n \in \mathbb{Q}$ is a p-adic Cauchy sequence, then by Exercise 4.38 there is an r such that $p^r x_n \in \mathbb{Z}_{(p)}$ for all n. This implies $p^r\underline{x} \in \mathbb{Z}_p$, where $\underline{x}$ is the p-adic number represented by the sequence x_n. This shows that the p-adic numbers have the form $p^{-r}\underline{y}$, where $\underline{y} \in \mathbb{Z}_p$. By Proposition 4.5, p-adic numbers may be written uniquely in the form,

$$\underline{x} = \sum_{n=-r}^{\infty} a_n p^n,$$

with coefficients $a_n \in \{0, 1, \ldots, p-1\}$.

In fact, $\mathbb{Q}_p$ is a field. To see this, note that 0 and 1 are not the same element of $\mathbb{Q}_p$, because they are not congruent modulo p. To show that a non-zero element has an inverse, suppose $\underline{x}$ is a non-zero p-adic number. There is a smallest integer a such that $p^a\underline{x} \in \mathbb{Z}_p$, and for this a we have $v_p(p^a\underline{x}) = 0$. By Proposition 4.9, $p^a\underline{x}$ has an inverse $\underline{y}$ in $\mathbb{Z}_p$; hence, $p^a\underline{y}$ is an inverse of $\underline{x}$.

Exercise 4.36. Show that the ultrametric inequality implies the triangle inequality $|x+y|_p \leq |x|_p + |y|_p$.

Exercise 4.37. Show that equivalence of p-adic Cauchy sequences is an equivalence relation. Show that if x_n and y_n are p-adic Cauchy sequences, then $(x_n + y_n)$ and $(x_n \cdot y_n)$ are p-adic Cauchy sequences. Show that if x_n is equivalent to x_n' and y_n is equivalent to y_n', then $x_n + y_n$ is equivalent to $x_n' + y_n'$ and $x_n \cdot y_n$ is equivalent to $x_n' \cdot y_n'$.

Exercise 4.38. Let $x_n \in \mathbb{Q}$ be a p-adic Cauchy sequence. Show that there exists an $r \in \mathbb{Z}$ such that $p^r x_n \in \mathbb{Z}_{(p)}$ for all n.

Exercise 4.39. Show that $\mathbb{Z}_p$ is an uncountable set.

HINTS FOR SOME EXERCISES

4.5. If $p \equiv 1 \bmod 7$, then by looking at the proof of the existence of primitive roots, show that Φ_7 has a root in $\mathbb{F}_p$. Then show that this root satisfies the conditions of Hensel's Lemma. If $p \not\equiv 1 \bmod 7$, then show that Φ_7 has no roots modulo p, using the corollary to Lagrange's Theorem.

4.7. In the second part, show that there are polynomials $h, k \in \mathbb{Z}_{(p)}[X]$ and an integer $a \geq 0$ such that $h(X)f(X) + k(X)f'(X) = p^a$.

4.24. In part 2, note that the series is $-\log(-1)$. To prove the equation in part 4, differentiate both sides of the equation.

4.31. Use the isomorphism $(\mathbb{Z}/p^n)^\times \cong \mathbb{F}_p^\times \times p\mathbb{Z}/p^n$.

4.35. Suppose the root in $\mathbb{Z}[\zeta_p]$ is $\sum a_i {\zeta_p}^i$. Take $x = \sum a_i g^{(q-1)i/p}$, where g is a Teichmüller lift of a primitive root modulo q.

Chapter 5

Diophantine Equations and Quadratic Rings

5.1 DIOPHANTINE EQUATIONS AND UNIQUE FACTORIZATION

A *Diophantine equation* is an equation of the form,

$$f(x_1, \ldots, x_n) = 0,$$

where f is a polynomial with integer coefficients. Number theorists are typically interested in finding solutions in integers to such equations.

There is no general method for solving Diophantine equations. Indeed, it is proven that no algorithm exists, which determines whether or not a general Diophantine equation has a solution in integers (see Part II, Chapter 3 in Introduction to Modern Number Theory: Fundamental Problems, Ideas and Theories [15]). Rather than looking at all Diophantine equations, we'll just look at a few examples. First, consider the equation

$$x^3 = 2y^2 + 2.$$

We'll show using congruences that there are no solutions in integers. Indeed, if (x, y) were a solutions then x would be even, and so x^3 would be a multiple of 8. This would imply that $y^2 + 1$ is a multiple of 4. However, we can easily check that y^2 is never congruent to -1 modulo 4. Therefore, there are no solutions.

Let's consider the following less trivial example.

$$x^3 = 2y^2 - 2.$$

This equation does have integer solutions. For example, $(x, y) = (0, 1)$ and $(0, -1)$ are solutions. We'll try to find all the solutions. If (x, y) is any solution, then just as in the previous example one sees that x is even, and so $y^2 - 1$ is a multiple of 4. This implies that y is odd. To simplify matters we make the substitutions $x = 2r$, $y = 2s+1$. In terms of the new variables r and s, we have

$$r^3 = s(s+1).$$

The integers s and $s+1$ are coprime, and their product is a perfect cube. From this, it follows that both s and $s+1$ are cubes. This can only happen if $s = 0$ or $s = -1$. These two possibilities give us $y = 1$ or $y = -1$, and hence $x = 0$. This shows that the obvious solutions $(0, 1)$ and $(0, -1)$ are the only solutions in integers. In this argument, we used the following lemma; we state the lemma for a unique factorization domain (UFD), although we have so far only used it with the ring $\mathbb{Z}$.

Lemma 5.1. *Let R be a UFD. Suppose a, b and c are elements of R such that a and b are coprime and $a \cdot b = c^n$. Then there are units $u, v \in R^\times$ and elements $d, e \in R$, such that*

$$a = ud^n, \quad b = ve^n.$$

Before proving the lemma, we'll consider another equation:

$$x^3 = y^2 + 1.$$

One can find an obvious solution $(1, 0)$ to this equation, so it's clear that no congruence argument can be used to bound the number of solutions. We'd like to try the same approach as in the previous example. This would involve factorizing the polynomial $y^2 + 1$. Unfortunately, this polynomial is irreducible over $\mathbb{Z}$. One way in which we can proceed is to factorize $y^2 + 1$ over a bigger ring. Over the ring $\mathbb{Z}[i]$ of Gaussian integers, we have

$$x^3 = (y+i)(y-i).$$

A little later in this chapter, we'll see that the ring $\mathbb{Z}[i]$ is a UFD, and so we may use Lemma 5.1 just as we did before. In order to use this lemma, we must show that the factors $y+i$ and $y-i$ are coprime in $\mathbb{Z}[i]$. Suppose $D \in \mathbb{Z}[i]$ is a common factor of $y+i$ and $y-i$. Since $2i = (y+i)-(y-i)$, it follows that D is a factor of $2i$, and hence of 2. Also, D is a factor of x. The integer x must be odd by a congruence argument modulo 4. By choosing integers h such that k such that $2h + xk = 1$, we see that D is a factor of 1.

This shows that $y+i$ and $y-i$ are coprime in $\mathbb{Z}[i]$. By Lemma 5.1, there is a unit $u \in \mathbb{Z}[i]$ such that

$$y+i = u(r+si)^3, \quad r,s \in \mathbb{Z}.$$

The only units in the ring $\mathbb{Z}[i]$ are the numbers $1, -1, i$ and $-i$. Each of these units is a cube in $\mathbb{Z}[i]$, so without loss of generality, we may assume that $u=1$. Expanding out our equation, we get

$$y+i = \left(r^3 - 3rs^2\right) + i\left(3r^2s - s^3\right).$$

Equating the real and imaginary parts, we get two simultaneous equations:

$$y = r^3 - 3rs^2 \quad \text{and} \quad 1 = 3r^2s - s^3.$$

The second of these equations shows that s is a factor of 1, and therefore $s = \pm 1$. The case $s = 1$ has no integer solution r, but if we set $s = -1$ then $r = 0$ is a solution. Substituting $r = 0$ and $s = -1$ into the first of the two simultaneous equations, we find $y = 0$. From the original equation $x^3 = y^2 + 1$, we have $x = 1$. Therefore, the solution $(1,0)$ which we found at the start is the only solution.

In this last example, we have used some properties of the ring $\mathbb{Z}[i]$ which we have not yet proved. We have use the fact that $\mathbb{Z}[i]$ has unique factorization. We have also used the fact that the only units in $\mathbb{Z}[i]$ are ± 1 and $\pm i$. In general, given a Diophantine equation of the form $y^n = f(x)$, we can try to solve it by the same method. One can always factorize the polynomial f over a suitable ring of algebraic integers $\mathbb{Z}[\alpha]$, but one is then faced with two questions:

(1) Does the ring $\mathbb{Z}[\alpha]$ have unique factorization?
(2) What are the units in $\mathbb{Z}[\alpha]$?

In this chapter, we'll examine these questions in the case that f is a quadratic polynomial. The same methods work more generally, but become more complicated as the degree of f increases.

Proof of Lemma 5.1. If $c = 0$, then without loss of generality, we can assume $a = 0$. Since a and b are coprime, it follows that b is a unit. Hence, the lemma is trivially true in this case.

Assume now that $c \neq 0$. We may factorize c as follows:

$$c = UP_1 \cdots P_r,$$

with U a unit and each P_i irreducible. We'll prove the lemma by induction on r. If $r = 0$, then a and b must both be units and the result is true.

Assume the result is true for $r-1$. We have

$$a \cdot b = U^n P_1^n \cdots P_r^n.$$

By uniqueness of factorization, we know that P_r is a factor of either a or b. Without loss of generality, we'll assume P_r is a factor of a. Since a and b are coprime, we know that P_r is not a factor of b, and therefore $a = P_r^n a'$ for some $a' \in R$. This implies $a' \cdot b = c'^n$, where $c' = UP_1 \cdots P_{r-1}$. By the inductive hypothesis, there exist units u, v and elements d, e in R such that $a' = ud^n$ and $b = ve^n$. Therefore, $a = u(dP_r)^n$. □

Exercise 5.1. Find all integer solutions to each of the following Diophantine equations.

$$y^6 = 7x^3 + 2, \quad y^5 = (31x+30)(31x-1), \quad y^2 = x^3 + x.$$

Exercise 5.2. Show that if $f, g \in \mathbb{C}[X]$ satisfy the equation $f^3 = g^2 + g$, then f and g are constant.

Exercise 5.3. Find all polynomials $f, g \in \mathbb{F}_3[X]$ such that $f^3 = g^2 + Xg$.

Exercise 5.4. Find all polynomials $f, g \in \mathbb{C}[X]$ such that $f^3 = g^2 + X$.

5.2 QUADRATIC RINGS

An integer d is called *square-free* if d is not a multiple of any square apart from 1^2, or equivalently if d is not a multiple of a square of a prime. Let d be a square-free integer and assume $d \neq 0, 1$. For each such d, we define a complex number $\alpha = \alpha_d$ as follows:

$$\alpha_d = \begin{cases} \sqrt{d} & \text{if } d \not\equiv 1 \bmod 4, \\ \dfrac{1+\sqrt{d}}{2} & \text{if } d \equiv 1 \bmod 4. \end{cases}$$

If $d \not\equiv 1 \bmod 4$, then $\alpha = \sqrt{d}$, which is a root of the polynomial $X^2 - d$. In the other case where $d \equiv 1 \bmod 4$, we have $(\alpha - \frac{1}{2})^2 = \frac{d}{4}$. This implies

$$m(\alpha) = 0, \quad \text{where } m(X) = X^2 - X + \frac{1-d}{4}.$$

Since $\frac{1-d}{4}$ is an integer, the polynomial m has integer coefficients. In either case, we see that α is a root of a monic irreducible polynomial over $\mathbb{Z}$ of degree 2. Therefore, α is an algebraic integer, and by Corollary 3.6 the following set is a subring of $\mathbb{C}$:

$$\mathbb{Z}[\alpha] = \{x + y\alpha : x, y \in \mathbb{Z}\}.$$

We call the ring $\mathbb{Z}[\alpha]$ a *quadratic ring*.[a] If $d > 0$, then the elements of $\mathbb{Z}[\alpha]$ are all real numbers, and we call $\mathbb{Z}[\alpha]$ a *real quadratic ring*. If $d < 0$, then we call $\mathbb{Z}[\alpha]$ a *complex quadratic ring*.

For example, consider the case $d = -1$. Since $-1 \not\equiv 1 \bmod 4$, we have $\alpha = \sqrt{-1}$. In this case, the quadratic ring is $\mathbb{Z}[i]$, the ring of of *Gaussian integers*.

There is another example which we have also seen before. Consider the case $d = -3$. Since $-3 \equiv 1 \bmod 4$, we have $\alpha = \frac{1+\sqrt{-3}}{2}$. Note that $\alpha = e^{\frac{2\pi i}{6}}$ and $\alpha - 1 = e^{\frac{2\pi i}{3}}$. In this case, the quadratic ring is the same as the cyclotomic ring $\mathbb{Z}[\zeta_3]$, which we used in the proof of the Quadratic Reciprocity Law. This particular ring is often called the ring of *Eisenstein integers*. Some elements are shown in Figure 5.1. The elements are the vertices of regular triangles which tessellate the plane.

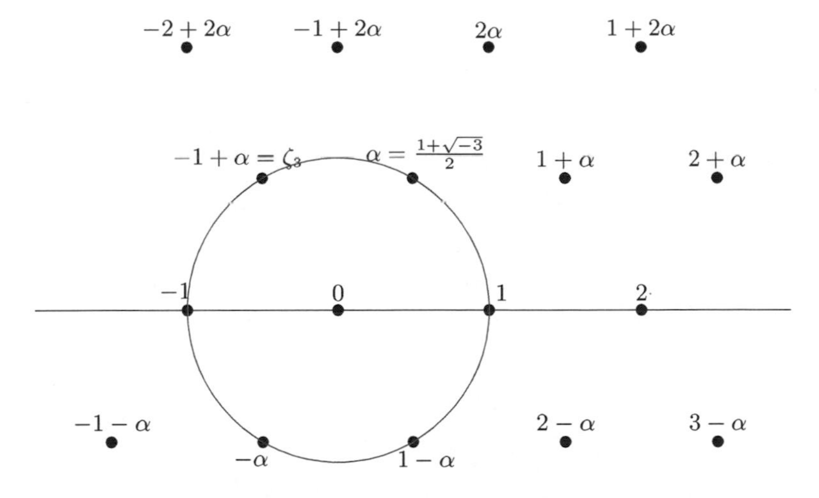

Figure 5.1: The Eisenstein integers.

[a] It is more usual to call these rings "the ring of algebraic integers in a quadratic field", but this is less practical here.

Every quadratic ring $\mathbb{Z}[\alpha_d]$ is contained in the field:

$$\mathbb{Q}(\sqrt{d}) = \{x + y\sqrt{d} : x, y \in \mathbb{Q}\}.$$

To see that $\mathbb{Q}(\sqrt{d})$ is indeed a field, note that it certainly contains 0 and 1 and is closed under addition, subtraction and multiplication so it is a ring. To see that every non-zero element has an inverse, note that if $x + y\sqrt{d} \neq 0$ then we have

$$\frac{1}{x + y\sqrt{d}} = \frac{1}{x^2 - dy^2}\left(x - y\sqrt{d}\right).$$

The denominator $x^2 - dy^2$ is non-zero because $\sqrt{d}$ is irrational.

At first sight, it seems odd to focus on the ring $\mathbb{Z}[\frac{1+\sqrt{5}}{2}]$ rather than the ring $\mathbb{Z}[\sqrt{5}]$ or any other subring of $\mathbb{Q}(\sqrt{5})$. The reason is given by the following proposition.

Proposition 5.1. *Let R be a subring of $\mathbb{C}$ and assume R is a UFD. If there are elements $x, y \in R$ such that $\frac{x}{y}$ is an algebraic integer, then $\frac{x}{y} \in R$.*

The number $\frac{1+\sqrt{5}}{2}$ is an algebraic integer, and is a fraction of two elements of $\mathbb{Z}[\sqrt{5}]$. Therefore, by Proposition 5.1, the ring $\mathbb{Z}[\sqrt{5}]$ stands no chance at all of having unique factorization. In contrast, we'll see later that the slightly bigger ring $\mathbb{Z}[\frac{1+\sqrt{5}}{2}]$ does have unique factorization. This argument shows that if R is any UFD containing $\sqrt{d}$ then R must contain α_d, and therefore R must contain the quadratic ring $\mathbb{Z}[\alpha_d]$. In fact, we can show that $\mathbb{Z}[\alpha_d]$ is exactly the set of algebraic integers in $\mathbb{Q}(\sqrt{d})$ (see Exercise 5.7).

Proof of Proposition 5.1. Without loss of generality, we can assume that x and y have no common factor in R. To show that $\frac{x}{y} \in R$, we'll show that y is a unit in R, or equivalently that y has no irreducible factors in R. Since $\frac{x}{y}$ is an algebraic integer, there is a polynomial equation of the form,

$$\left(\frac{x}{y}\right)^n + a_{n-1}\left(\frac{x}{y}\right)^{n-1} + \cdots + a_1\left(\frac{x}{y}\right) + a_0 = 0, \quad a_0, \ldots, a_{n-1} \in \mathbb{Z}.$$

Multiplying through by y^n, we get

$$x^n + a_{n-1}x^{n-1}y + \cdots + a_1 x y^{n-1} + a_0 y^n = 0.$$

Suppose P is an irreducible factor of y. From the equation, we see that P is a factor of x^n. Since R has unique factorization, it follows that P is a factor of x, which contradicts the assumption that x and y are coprime. Therefore, y has no irreducible factors so y is a unit. □

Conjugates, norms and units. The *conjugate* of an element of $\mathbb{Q}(\sqrt{d})$ is defined by

$$\overline{x + y\sqrt{d}} = x - y\sqrt{d}.$$

If $d \not\equiv 1 \bmod 4$, then $\alpha = \sqrt{d}$ so we have $\bar{\alpha} = -\alpha$. On the other hand, if $d \equiv 1 \bmod 4$ then we have $\alpha = \frac{1+\sqrt{d}}{2}$, and therefore

$$\bar{\alpha} = \tfrac{1-\sqrt{d}}{2} = 1 - \alpha.$$

Lemma 5.2 (Properties of conjugates). *For all elements $A, B \in \mathbb{Q}(\sqrt{d})$, we have*

(1) $\overline{A \cdot B} = \overline{A} \cdot \overline{B}$, $\overline{A + B} = \overline{A} + \overline{B}$ *and* $\overline{\overline{A}} = A$;
(2) *if* $A \in \mathbb{Z}[\alpha]$ *then* $\overline{A} \in \mathbb{Z}[\alpha]$.

Proof. This is left as Exercise 5.5. □

Definition. For an element $A \in \mathbb{Q}(\sqrt{d})$, the *norm* of A is defined by

$$\mathrm{N}(A) = A\bar{A}.$$

As elements have the form $x + y\sqrt{d}$, we have

$$\mathrm{N}(x + y\sqrt{d}) = (x + y\sqrt{d})(x - y\sqrt{d}) = x^2 - d \cdot y^2.$$

If $d \equiv 1 \bmod 4$, then we have $\alpha = \frac{1+\sqrt{d}}{2}$. In this case, we have:

$$\left(x + y\frac{1+\sqrt{d}}{2}\right) = \left(x + \frac{y}{2}\right)^2 - d\left(\frac{y}{2}\right)^2 = x^2 + xy + \left(\frac{1-d}{4}\right)y^2.$$

Taking these two cases together, we have

$$\mathrm{N}(x + y\alpha) = \begin{cases} x^2 - dy^2 & \text{if } d \not\equiv 1 \bmod 4, \\ x^2 + xy + \left(\dfrac{1-d}{4}\right)y^2 & \text{if } d \equiv 1 \bmod 4. \end{cases} \tag{5.1}$$

In either case, we see that the norm of an element of $\mathbb{Z}[\alpha]$ is an integer, and the norm of an element of $\mathbb{Q}(\sqrt{d})$ is a rational number.

If $\mathbb{Z}[\alpha]$ is a complex quadratic ring (i.e. $d < 0$), then the conjugate of an element is just its complex conjugate. This implies $\mathrm{N}(A) = |A|^2$, where $|A|$ is the absolute value of the complex number A. In particular, $\mathrm{N}(A) \geq 0$. On the other hand, if $\mathbb{Z}[\alpha]$ is a real quadratic ring then $\bar{A}$ is not the complex conjugate of A, and $\mathrm{N}(A)$ takes both positive and negative values.

Proposition 5.2. *For all* $A, B \in \mathbb{Q}(\sqrt{d})$, *we have* $\mathrm{N}(A \cdot B) = \mathrm{N}(A) \cdot \mathrm{N}(B)$. *If* $\mathrm{N}(A) = 0$, *then* $A = 0$.

Proof. The proof is left as Exercise 5.5. □

Recall that a *unit* is an element of a ring with a multiplicative inverse. Such elements are also called the *invertible elements* in the ring. The set of units in a ring R is written $R^{\times}$, and this set is a group with the operation of multiplication (see Proposition 1.1).

Corollary 5.1. *An element* $A \in \mathbb{Z}[\alpha]$ *is a unit if and only if* $\mathrm{N}(A) = \pm 1$.

Proof. If $\mathrm{N}(A) = \pm 1$, then $A\bar{A} = \pm 1$ and so $A^{-1} = \pm \bar{A}$, which is an element of $\mathbb{Z}[\alpha]$. Conversely if A is invertible then

$$\mathrm{N}(A) \cdot \mathrm{N}(A^{-1}) = \mathrm{N}(A \cdot A^{-1}) = \mathrm{N}(1) = 1.$$

Therefore, $\mathrm{N}(A)$ is a factor of 1, so $\mathrm{N}(A) = \pm 1$. □

In the case of a complex quadratic ring, the corollary shows that A is a unit if and only if $|A| = 1$, so the units are the elements of the ring which are on the unit circle in $\mathbb{C}$. As an example, we'll find the units in the ring $\mathbb{Z}[i]$ of Gaussian integers. We have $\overline{x + yi} = x - yi$. Therefore,

$$\mathrm{N}(x + yi) = (x + yi)(x - yi) = x^2 + y^2.$$

To find the units, we need to find all integer solutions to $x^2 + y^2 = 1$. These solutions are $x = \pm 1$, $y = 0$ and $x = 0$, $y = \pm 1$. Therefore, the units in $\mathbb{Z}[i]$ are 1, -1, i and $-i$.

As another example, we'll find all the units in the ring of Eisenstein integers. In this case, $\alpha = \frac{1+\sqrt{-3}}{2} = e^{\frac{2\pi i}{6}}$, and by Equation (5.1) we have

$$\mathrm{N}(x + y\alpha) = x^2 + xy + y^2.$$

To find the units, we must find all integer solutions to $x^2 + xy + y^2 = 1$. Completing the square in this equation, we get

$$\left(x + \frac{1}{2}y\right)^2 + \frac{3}{4}y^2 = 1.$$

It follows that $y^2 \leq \frac{4}{3}$. Since y is an integer, this implies $|y| \leq 1$. Similarly $|x| < 1$. If $y = 0$, then we must have $x = \pm 1$, and if $y = \pm 1$ then $x^2 + xy + 1 = 1$, so $x = 0, -y$ are both solutions. This gives us the six solutions:

$$(1, 0),\ (-1, 0),\ (0, 1),\ (-1, 1),\ (0, -1),\ (1, -1).$$

Each solution corresponds to a unit, so the units in the Eisenstein integers are:

$$1, \ -1, \ \alpha, \ -1+\alpha, \ -\alpha, \ 1-\alpha.$$

Equivalently, these numbers are ± 1, $\pm\zeta_3$ and $\pm\zeta_3^2$. We can see these units on the unit circle in $\mathbb{C}$ in Figure 5.1.

Generalizing these two examples, we have a complete description of all units in complex quadratic rings:

Theorem 5.1. *Let d be a negative square-free integer and let $\mathbb{Z}[\alpha]$ be the corresponding complex quadratic ring. Then the units in $\mathbb{Z}[\alpha]$ are as follows:*

$$\mathbb{Z}[\alpha]^\times = \begin{cases} \{1, -1, i, -i\} & \textit{if } d = -1, \\ \{1, -1, \alpha, -\alpha, \alpha - 1, 1 - \alpha\} & \textit{if } d = -3, \\ \{1, -1\} & \textit{in all other cases.} \end{cases}$$

Proof. We've already proved the theorem in the special cases $d = -1$ and $d = -3$, so we can assume d is neither -1 nor -3. Assume first that $d \not\equiv 1 \bmod 4$. Then we have $\mathrm{N}(x + y\alpha) = x^2 - dy^2$, so we need to find the integer solutions to the equation,

$$x^2 - dy^2 = 1.$$

Note that $-d > 1$, so we must have $y = 0$, and hence $x = \pm 1$. The two solutions $(1, 0)$, $(-1, 0)$ give the two units $1, -1$.

Assume now that $d \equiv 1 \bmod 4$. Since we are ruling out the case $d = -3$, it follows that $d \leq -7$. We need to find the solutions to the equation,

$$x^2 + xy + \frac{1-d}{4}y^2 = 1.$$

Completing the square, we get

$$\left(x + \frac{1}{2}y\right)^2 - \frac{d}{4}y^2 = 1.$$

Since $-\frac{d}{4} > 1$, we have $y = 0$. This implies $x = \pm 1$ as in the first case. $\square$

We've found all the units in each complex quadratic ring. In each case, the group $\mathbb{Z}[\alpha]^\times$ is finite. We'll see in Section 5.6 that each real quadratic ring has infinitely many invertible elements. Consider for example the case $d = 5$, so $\alpha = \frac{1+\sqrt{5}}{2}$. Some of the units in the ring $\mathbb{Z}[\frac{1+\sqrt{5}}{2}]$ are shown in red in Figure 5.2. In this picture, a number $a + b\sqrt{5}$ is shown as the point $(a, b\sqrt{5})$ in the plane. The elements of norm 1 and -1 correspond to points (x, y) on the hyperbolae $x^2 - y^2 = \pm 1$. These hyperbolae are shown in

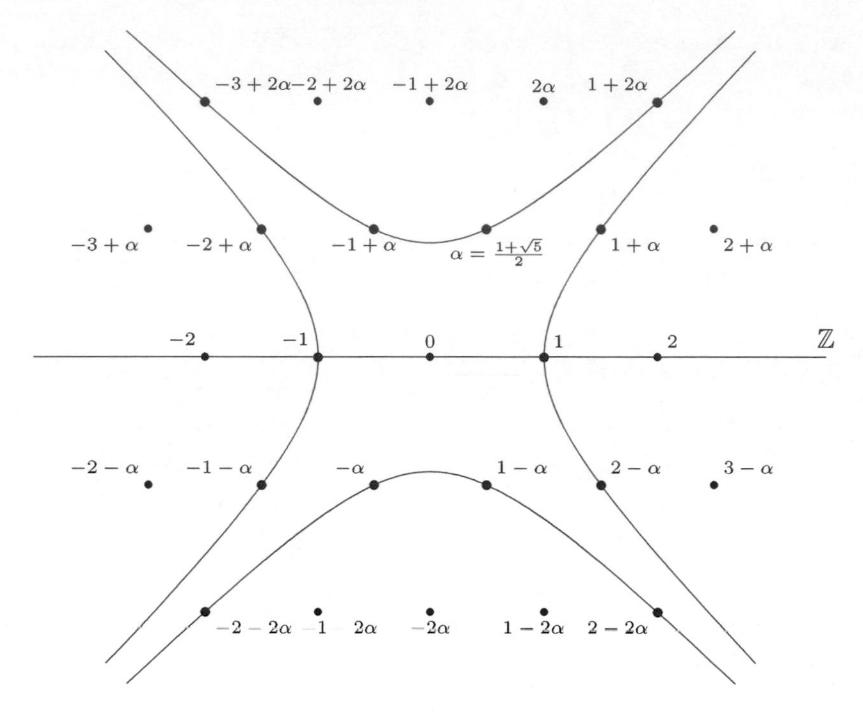

Figure 5.2: Units in $\mathbb{Z}[\frac{1+\sqrt{5}}{2}]$.

red. In particular, we see that $\mathrm{N}(\frac{1+\sqrt{5}}{2}) = -1$, so the element $\alpha = \frac{1+\sqrt{5}}{2}$ is invertible. To see that there are infinitely many invertible elements, note that the powers α^n are invertible for all $n \in \mathbb{Z}$. No two of these powers are equal, because the sequence of real numbers α^n is increasing.

Exercise 5.5. Prove Lemma 5.2 and Proposition 5.2.

Exercise 5.6. For an element $A = r + s\sqrt{d} \in \mathbb{Q}(\sqrt{d})$, find a matrix M such that

$$A(x + y\sqrt{d}) = x' + y'\sqrt{d}, \quad \text{where } \begin{pmatrix} x' \\ y' \end{pmatrix} = M \begin{pmatrix} x' \\ y' \end{pmatrix}.$$

Show that $\mathrm{N}(A) = \det M$. Show that $A \in \mathbb{Z}[\alpha]$ if and only if the characteristic polynomial $\det(XI_2 - M)$ has integer coefficients.

Exercise 5.7. Show that every element of $\mathbb{Z}[\alpha]$ is an algebraic integer. Conversely, show that if $m(X) = X^2 + bX + c$ with integers b and c, then the roots of m are in one of the quadratic rings.

5.3 NORM-EUCLIDEAN QUADRATIC RINGS

A quadratic ring $\mathbb{Z}[\alpha]$ is said to be *norm-Euclidean* if for every $A, B \in \mathbb{Z}[\alpha]$ with $B \neq 0$, there exist $Q, R \in \mathbb{Z}[\alpha]$ such that

$$A = QB + R \quad \text{and} \quad |\mathrm{N}(R)| < |\mathrm{N}(B)|.$$

In particular, a norm-Euclidean ring $\mathbb{Z}[\alpha]$ is a Euclidean domain in the sense of Section 2.5. Proposition 2.7 shows that if $\mathbb{Z}[\alpha]$ is norm-Euclidean then $\mathbb{Z}[\alpha]$ is a UFD. We'll see in this section that several of the quadratic rings are norm-Euclidean, and therefore have unique factorization.

As an example, we'll begin by showing that the ring $\mathbb{Z}[i]$ of Gaussian integers is norm-Euclidean. To see why this is the case, we note that the complex plane may be covered by squares of side 1, with each square centered on a Gaussian integer (see Figure 5.3). By Pythagoras' Theorem, the distance from the center of one of these squares to any other point in the square is no more than $\frac{1}{\sqrt{2}}$ (see Figure 5.4). This means that for any complex number Z there is a Gaussian integer Q, such that $|Q - Z| \leq \frac{1}{\sqrt{2}}$. To see how the division algorithm works, we'll see how to divide A by B with remainder. We first define the quotient Q to be the nearest Gaussian integer to $\frac{A}{B}$. This means Q is the center of the square containing $\frac{A}{B}$. As we saw, this implies $\left|\frac{A}{B} - Q\right| \leq \frac{1}{\sqrt{2}}$. We can rewrite this inequality as $|A - QB| \leq \frac{1}{\sqrt{2}}|B|$.

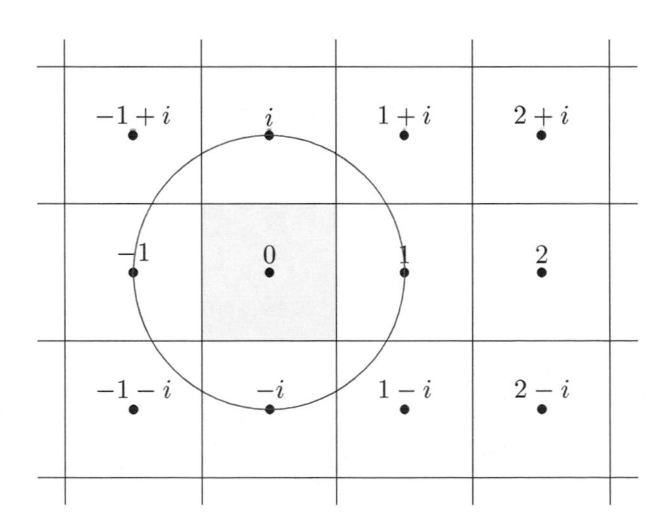

Figure 5.3: The norm-Euclidean property of the Gaussian integers.

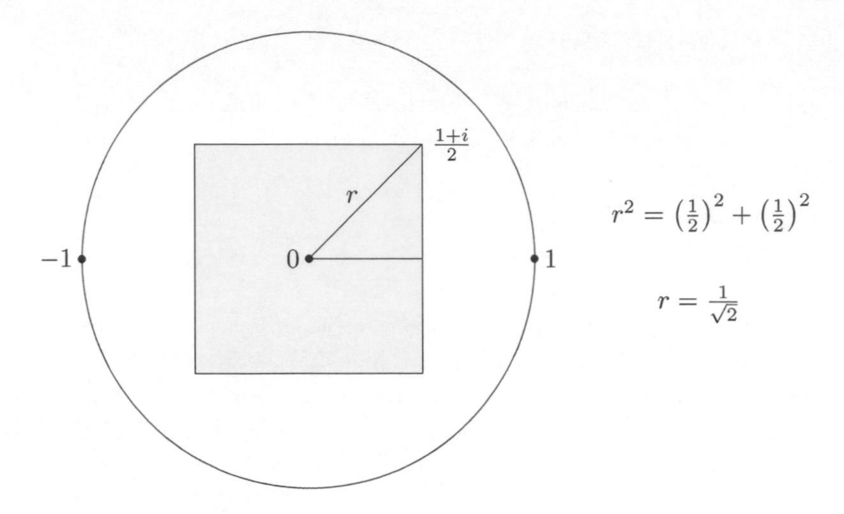

Figure 5.4: Square of side 1 centered at 0.

Squaring this, we get

$$\mathrm{N}\left(A - QB\right) \leq \frac{1}{2} \cdot \mathrm{N}(B). \tag{5.2}$$

Having chosen the quotient Q, the remainder must be given by $R = A - QB$. We see that both Q and R are Gaussian integers and we have $A = QB + R$. Furthermore, Equation (5.2) implies that $\mathrm{N}(R) < \mathrm{N}(B)$. Therefore, $\mathbb{Z}[i]$ is norm-Euclidean.

To see how this works in practice, let $A = 34 + 6i$ and $B = 7 + 3i$. We have

$$\frac{A}{B} = \frac{128}{29} - \frac{30}{29}i.$$

The nearest Gaussian integer to $\frac{A}{B}$ is $4 - i$, so we set $Q = 4 - i$ and $R = A - QB = 3 + i$. The remainder R has norm 10 and $\mathrm{N}(B) = 58$, so $\mathrm{N}(R) < \mathrm{N}(B)$.

The key point in the discussion above is that every element in the square centered at 0 has norm less than 1. In other words, the square of side 1 fits inside a circle of radius 1 (see Figure 5.4). This same argument works for several other quadratic rings, although the square of side 1 and the circle of radius 1 are replaced by different shapes.

Example. Let $d = -7$, so $\alpha = \frac{1+\sqrt{-7}}{2}$. We'll show that the quadratic ring $\mathbb{Z}[\alpha]$ is norm-Euclidean. To prove this, we copy roughly the argument used for the Gaussian integers. We can cover the complex plane with rectangles centered on the elements of $\mathbb{Z}[\alpha]$. These rectangles have width 1 and height $\frac{\sqrt{7}}{2}$. If we write r for the furthest distance from a point on the rectangle to the center of the rectangle, then by Pythagoras' Theorem we have $r^2 = \frac{11}{16}$, and in particular $r < 1$. Therefore, all points in the rectangle are contained in the circle of radius 1 with the same center. Again, this implies that $\mathbb{Z}[\alpha]$ is norm-Euclidean. The argument is illustrated in Figure 5.5.

The examples above show the limitations of this method. If we look at the ring $\mathbb{Z}[\sqrt{-5}]$ instead of $\mathbb{Z}[\frac{1+\sqrt{-7}}{2}]$, then the height of the yellow rectangle in this case is $\sqrt{5}$, which is larger than 2. Therefore, the rectangle is not contained in the red circle of radius 1 (see Figure 5.6). There will be complex numbers whose distance from every element of $\mathbb{Z}[\sqrt{-5}]$ is bigger than 1. If we choose A and B in $\mathbb{Z}[\sqrt{-5}]$ so that $\frac{A}{B}$ is such a number, then no matter what element $Q \in \mathbb{Z}[\sqrt{-5}]$ we choose, the remainder $R = A - QB$ will always have norm bigger than B (see Exercise 5.9 for an example). Therefore, $\mathbb{Z}[\sqrt{-5}]$ is not norm-Euclidean.

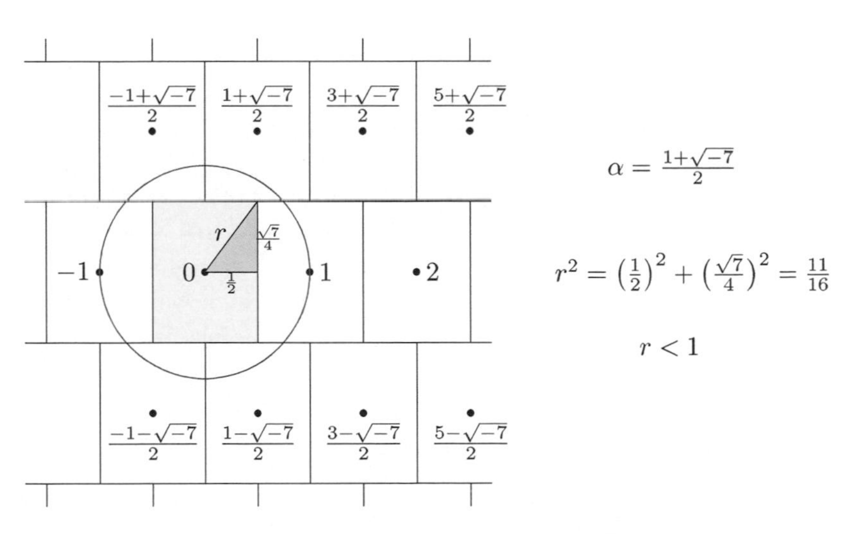

Figure 5.5: The ring $\mathbb{Z}[\frac{1+\sqrt{-7}}{2}]$ is norm-Euclidean.

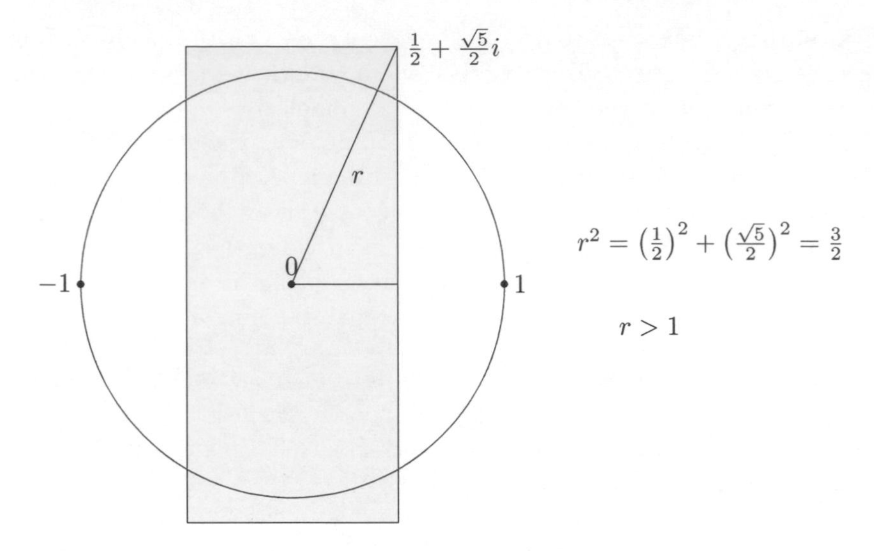

Figure 5.6: $\mathbb{Z}[\sqrt{-5}]$ is not norm-Euclidean.

Example. The picture for real quadratic rings is rather different. We'll illustrate this with the example $d = 3$. In this case, the quadratic ring is $\mathbb{Z}[\sqrt{3}]$.

Since $\mathbb{Z}[\sqrt{3}]$ is contained in the real numbers, we'll need a different way of drawing pictures of the ring elements, instead of picturing them as points in the complex plane. Instead, we shall represent a number $x + y\sqrt{3}$ with $x, y \in \mathbb{Q}$ by the point $(x, y\sqrt{3})$ in the plane. Just as before, the plane $\mathbb{R}^2$ may be covered by rectangles centered on elements of $\mathbb{Z}[\sqrt{3}]$ with width 1 and height $\sqrt{3}$. The norm of an element at the point (x, y) is not $x^2 + y^2$, but is instead $x^2 - y^2$. Therefore, the points of $\mathbb{Q}(\sqrt{3})$ satisfying $|\mathrm{N}(Z)| \leq 1$ are enclosed by the hyperbolae $x^2 - y^2 = \pm 1$. These hyperbolae are shown in red in Figure 5.7. The crucial point to note in this picture, is that the yellow rectangle centered at $(0, 0)$ is contained in the region bounded by these hyperbolae. Therefore, elements of $\mathbb{Q}(\sqrt{3})$ in this rectangle satisfy $|\mathrm{N}(Z)| < 1$. Using this fact, we can show just as before that $\mathbb{Z}[\sqrt{3}]$ is norm-Euclidean.

Using the same method, one can prove the following:

Theorem 5.2 (The Disappointing Theorem). *The quadratic rings with* $d = -1, -2, -3, -7, -11, 2, 3, 5,$ *and* 13 *are norm-Euclidean.*

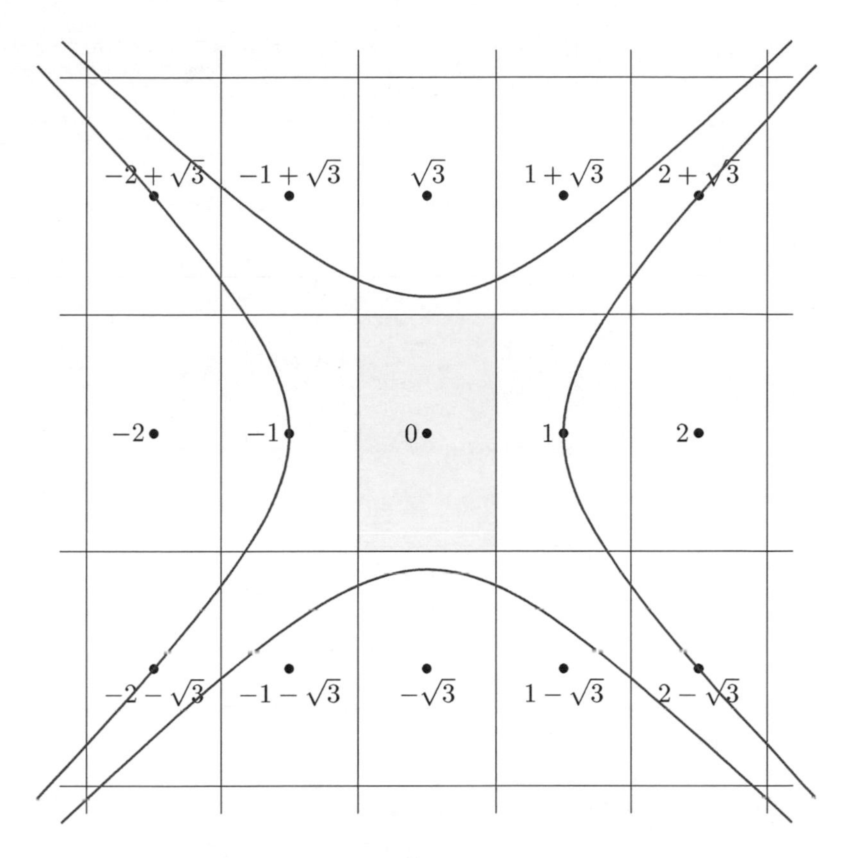

Figure 5.7: $\mathbb{Z}[\sqrt{3}]$ is norm-Euclidean.

By working considerably harder, one can show that $\mathbb{Z}[\alpha]$ is norm-Euclidean in the following additional cases: $d = 6, 7, 11, 17, 19, 21, 29, 33, 37, 41, 57, 73$ (see Exercise 5.11 for an example of how to prove this). To prove the theorem, we'll use the following lemma.

Lemma 5.3. *Suppose $Z \in \mathbb{Q}(\sqrt{d})$. Then there is a $Q \in \mathbb{Z}[\alpha]$ such that if $Z - Q = x + y\sqrt{d}$, then*

(1) $|x| \leq \frac{1}{2}$,
(2) *If* $d \not\equiv 1 \bmod 4$, *then* $|y| \leq \frac{1}{2}$,
(3) *If* $d \equiv 1 \bmod 4$, *then* $|y| \leq \frac{1}{4}$.

In the cases $d = -1, -2, -3, -7, -11, 2, 3, 5, 13$, *we have* $|\mathrm{N}(Z - Q)| < 1$.

Proof of the Disappointing Theorem. Let $A, B \in \mathbb{Z}[\alpha]$ with $B \neq 0$. By Lemma 5.3, we can find a $Q \in \mathbb{Z}[\alpha]$ such that $\left|\mathrm{N}\left(\frac{A}{B} - Q\right)\right| < 1$. Let $R = A - BQ$. Then clearly $A = QB + R$ and we have $|\mathrm{N}(R)| = \left|\mathrm{N}\left(\frac{A}{B} - Q\right)\right| \cdot |\mathrm{N}(B)| < |\mathrm{N}(B)|$. □

Proof of Lemma 5.3. Assume first $d \not\equiv 1 \bmod 4$ and let $Z = u + v\sqrt{d}$ with u and v in $\mathbb{Q}$. Then we let $Q = r + s\sqrt{d}$, where r and s are the nearest integers to u and v. In particular, $x = u - r$ and $y = v - s$ have absolute value $\leq \frac{1}{2}$. In the cases $d = -1, -2$, we have

$$0 \leq \mathrm{N}(Z - Q) = x^2 - dy^2 \leq \frac{1 + |d|}{4} < 1.$$

In the cases $d = 2, 3$, the norm $x^2 - dy^2$ is bounded by

$$-\frac{3}{4} \leq -dy^2 \leq x^2 - dy^2 \leq x^2 \leq \frac{1}{4}.$$

This shows that $|\mathrm{N}(Z - Q)| < 1$ in these cases as well.

Now assume $d \equiv 1 \bmod 4$ and let $Z = u + v\sqrt{d}$. Let s be the nearest integer to $2v$, so $|s - 2v| \leq \frac{1}{2}$. This implies $|\frac{s}{2} - v| \leq \frac{1}{4}$. Next, let r be the nearest integer to $u - \frac{s}{2}$. Hence, $|r + \frac{s}{2} - u| \leq \frac{1}{2}$. If we set $Q = r + s\alpha$, then $Z - Q = x + y\sqrt{d}$ with $|x| = |u - r - \frac{s}{2}| \leq \frac{1}{2}$ and $|y| = |v - \frac{s}{2}| \leq \frac{1}{4}$.

In the cases $d = -3, -7, -11$, we have

$$0 \leq \mathrm{N}(Z - Q) = x^2 - dy^2 \leq \frac{4 + |d|}{16} < 1.$$

In the real quadratic cases $d = 5, 13$, we have

$$-\frac{13}{16} \leq -\frac{d}{16} \leq -dy^2 \leq x^2 - dy^2 \leq x^2 \leq \frac{1}{4}.$$

In each of these cases, we have $|\mathrm{N}(Z - Q)| < 1$. □

An application to Diophantine equations. As an example of how to use the theorem above, we'll solve the following Diophantine equation:

$$x^3 = y^2 + 11. \tag{5.3}$$

Factorizing the right-hand side, we get

$$x^3 = (y + \sqrt{-11})(y - \sqrt{-11}).$$

The number $\sqrt{-11}$ is in the quadratic ring $\mathbb{Z}[\alpha]$, where $\alpha = \frac{1+\sqrt{-11}}{2}$; in fact $\sqrt{-11} = 2\alpha - 1$. We've shown in Theorem 5.2 that $\mathbb{Z}[\alpha]$ is norm-Euclidean, and therefore has unique factorization. We'll next show that the factors

$y + \sqrt{-11}$ and $y - \sqrt{-11}$ are coprime. Let $D \in \mathbb{Z}[\alpha]$ be a common factor of $y + \sqrt{-11}$ and $y - \sqrt{-11}$, and therefore of x. It follows that D is a factor of $2\sqrt{-11}$; hence D is a factor of 22. If x were even, then we would have $y^2 + 11 \equiv 0 \bmod 8$, which gives a contradiction. Therefore, x is odd. Similarly, if x is a multiple of 11 then $y^2 + 11 \equiv 0 \bmod 11^3$, which gives a contradiction. We've shown that x is coprime to 22, so there exist integers h and k such that $22h + xk = 1$. Since D is a common factor of x and 22, it follows that D is a factor of 1, and hence a unit. Therefore, $y + \sqrt{-11}$ and $y - \sqrt{-11}$ are coprime. By Lemma 5.1, we have

$$y + \sqrt{-11} = U(r + s\alpha)^3, \quad \text{for some } U \in \mathbb{Z}[\alpha]^\times, \quad \text{and} \quad r, s \in \mathbb{Z}.$$

By Theorem 5.1, we know that $U = \pm 1$. Since -1 is a cube, we may assume without loss of generality that $U = 1$. Expanding out our equation, we get

$$y - 1 + 2\alpha = r^3 + 3r^2 s\alpha + 3rs^2\alpha^2 + s^3\alpha^3.$$

Recall that α is a root of the polynomial $X^2 - X + 3$, so we have $\alpha^2 = \alpha - 3$. This implies $\alpha^3 = -2\alpha - 3$, and therefore

$$y - 1 + 2\alpha = (r^3 - 9rs^2 - 3s^3) + (3r^2 s + 3rs^2 - 2s^3)\alpha.$$

Equating the coefficients of 1 and α, we get simultaneous equations:

$$y - 1 = r^3 - 9rs^2 - 3s^3 \quad \text{and} \quad 3r^2 s + 3rs^2 - 2s^3 = 2. \tag{5.4}$$

The second equation shows that s is a factor of 2, so $s = \pm 1$ or ± 2. We'll consider each of these cases in turn. If $s = 1$, then the second of the simultaneous equations reduces to

$$3r^2 + 3r - 2 - 2.$$

This has no integer solutions. If $s = -1$, then we have instead

$$-3r^2 + 3r + 2 = 2.$$

This has solutions $r = 0$ and $r = 1$. Substituting these into Equation (5.4), we find that y is equal to 4 or -4. These values do indeed give solutions $(3, 4)$ and $(3, -4)$ to Equation (5.3). In the case $s = 2$, we have

$$6r^2 + 12r - 16 = 2.$$

This has solutions $r = 1$ and $r = -3$, which give $y = \pm 58$. We obtain two more solutions $(15, 58)$ and $(15, -58)$ to Equation (5.3). Finally in the case $s = -2$, we have

$$-6r^2 + 12r + 16 = 2,$$

which has no integer solutions. In summary, we've shown that Equation (5.3) has exactly four solutions in integers; they are $(3, 4)$, $(3, -4)$, $(15, 58)$ and $(15, -58)$.

Other quadratic rings with unique factorization. Theorem 5.2 gives several examples of norm-Euclidean quadratic rings. Such rings have unique factorization. There are also many examples of quadratic rings which have unique factorization, even though they are not norm-Euclidean. It's known that a complex quadratic ring has unique factorization for exactly the following values of d and no more:

$$d = -1, -2, -3, -7, -11, -19, -43, -67, -163.$$

This result was conjectured by Gauss in Article 303 of [10]. There are two proofs; one was found by Baker and is in the book [3]. The other proof was found by Stark based on earlier work of Heegner. Stark's proof is described in the book [5]. The history of the problem is described in [11]. Incidentally, it is relatively easy to show that $\mathbb{Z}[\alpha]$ has unique factorization in the cases listed above; the difficult part of the proof is to show that there are no other complex quadratic rings with unique factorization.

The first complex quadratic ring which is not in the list above, and thus does not have unique factorization, is the ring $\mathbb{Z}[\sqrt{-5}]$. It is not at all hard to find examples of non-unique factorization in this ring. For example,

$$6 = 2 \times 3 = (1 + \sqrt{-5}) \times (1 - \sqrt{-5}).$$

The elements 2, 3, and $1 \pm \sqrt{-5}$ are all irreducible in $\mathbb{Z}[\sqrt{-5}]$. To see this, note that these elements have norms 2, 9, 6 and 6, respectively. Hence, any proper factor would have norm 2 or 3. However, the ring $\mathbb{Z}[\sqrt{-5}]$ has no elements of norm 2 or 3 since $x^2 + 5y^2$ is never equal to 2 or 3 for integers x, y.

In contrast, it is much more common for a real quadratic ring to have unique factorization. The following conjecture of Gauss (see Article 304 of [10]) is believed, but has not been proved.

Conjecture. *There are infinitely many real quadratic rings with unique factorization.*

For example, $\mathbb{Z}[\alpha]$ has unique factorization for the following values of d (and many more):

$$d = 2, 3, 5, 6, 7, 11, 13, 14, 17, 19, 21, 22, 23, 29, 31, 33, 37, \ldots$$

Exercise 5.8. For each pair of elements A, B find Q and R such that $A = QB + R$ and $|\mathrm{N}(R)| < |\mathrm{N}(B)|$:

$$(18 + 5i, 7 + 3i), \quad (33 + 7\sqrt{2}, 4 + 3\sqrt{2}), \quad (40 + 6\alpha_5, 3 + 6\alpha_5).$$

Exercise 5.9. Let $A = \sqrt{-5}$ and $B = 2$. Show that if Q and R are in $\mathbb{Z}[\sqrt{-5}]$ with $A = QB + R$, then $\mathrm{N}(R) > \mathrm{N}(B)$. (This shows that $\mathbb{Z}[\sqrt{-5}]$ is not norm-Euclidean.)

Exercise 5.10. Find the highest common factor of $A = 11 - 13i$ and $B = 4 + 6i$ in $\mathbb{Z}[i]$, and find Gaussian integers H, K such that $\mathrm{hcf}(A, B) = HA + KB$.

Exercise 5.11. Let $Z = x + y\sqrt{6} \in \mathbb{Q}(\sqrt{6})$ with $|x|, |y| \leq \frac{1}{2}$. Show that either $|\mathrm{N}(Z)| < 1$ or $|\mathrm{N}(Z + 1)| < 1$ or $|\mathrm{N}(Z - 1)| < 1$. Hence, show that $\mathbb{Z}[\sqrt{6}]$ is norm-Euclidean.

Exercise 5.12. Find all solutions to each of the following Diophantine equations:

$$x^3 = y^2 + 2, \qquad x^3 = y^2 + y + 2, \qquad x^5 = y^2 + y + 2,$$

$$x^3 = y^2 + y + 3, \qquad x^5 = y^2 + y + 3, \qquad x^7 = y^2 + y + 5.$$

5.4 DECOMPOSING PRIMES IN QUADRATIC RINGS

In this section, we'll assume throughout that $\mathbb{Z}[\alpha]$ is a quadratic ring with unique factorization. This means that every non-zero element of this ring can be factorized uniquely into irreducible elements. Our next aim is to describe the irreducible elements in $\mathbb{Z}[\alpha]$.

Lemma 5.4. *Let $\mathbb{Z}[\alpha]$ be a quadratic ring with unique factorization. If Q is an irreducible element of $\mathbb{Z}[\alpha]$, then there is a unique prime number p in $\mathbb{Z}$, such that Q is a factor of p.*

Proof (*Existence*). The element Q certainly divides some integer n; for example, it divides its own norm $Q\bar{Q}$. On the other hand if $n = ab$, then by uniqueness of factorization, Q is a factor of either a or b. Hence, if we factorize n into primes in $\mathbb{Z}$, then one of these primes is a multiple of Q.

(*Uniqueness*). Suppose Q is a factor of two distinct prime numbers p and q. By Bezout's Lemma, we can find integers h and k such that $1 = hp + kq$. Hence, Q is a factor of 1, and is therefore a unit. This gives a contradiction since irreducible elements are not units. $\square$

Lemma 5.4 shows that in order to find the irreducible elements in $\mathbb{Z}[\alpha]$, we just need to factorize the primes of $\mathbb{Z}$ in the ring $\mathbb{Z}[\alpha]$. We'll now investigate the ways in which a prime number p can factorize. If $Q_1 \in \mathbb{Z}[\alpha]$ is an irreducible factor of a prime number p, then $\mathrm{N}(Q_1)$ must be a factor of $\mathrm{N}(p)$. Since $\mathrm{N}(p) = p^2$, it follows that $\mathrm{N}(Q_1)$ is either $\pm p$ or $\pm p^2$. In the case $\mathrm{N}(Q_1) = \pm p^2$, we must have $p = UQ_1$ for a unit U, and therefore p is itself irreducible. In this case, we say that p is *inert* in $\mathbb{Z}[\alpha]$.

In the case that Q_1 has norm $\pm p$, we have $p = Q_1Q_2$, where Q_2 also has norm $\pm p$. In this case, p is not irreducible in $\mathbb{Z}[\alpha]$. In fact, since $Q_1\bar{Q}_1 = \mathrm{N}(Q_1) = \pm p$, it follows that $Q_2 = \pm\bar{Q}_1$, and Q_2 is also irreducible. We'll distinguish between two different cases: either Q_2 is a unit multiple of Q_1, so we have $p = UQ_1^2$ for a unit U, or Q_1/Q_2 is not in $\mathbb{Z}[\alpha]$. In the case that $p = UQ_1^2$, we say that p is *ramified* in $\mathbb{Z}[\alpha]$; and when p has two distinct irreducible factors we say that p *splits* in $\mathbb{Z}[\alpha]$. To summarize, the prime p is called either inert, ramified or split, depending on which of the following three cases we are in:

(1) p is irreducible in $\mathbb{Z}[\alpha]$;
(2) $p = UQ^2$, where U is a unit and Q is irreducible with norm $\pm p$;
(3) $p = Q_1Q_2$, where Q_1 and Q_2 are irreducible elements with norm $\pm p$, and Q_1 is not a unit multiple of Q_2.

For example, consider the ring $\mathbb{Z}[i]$ of Gaussian integers. A prime p will factorize in this ring if there is an element with norm p. Indeed, if we have an element Q with norm p, then the factorization is $p = Q\bar{Q}$. Note that $\mathrm{N}(x+iy) = x^2+y^2$, so to check whether p factorizes we must check whether the equation $x^2+y^2 = p$ has solutions in integers. Here are some examples:

$$
\begin{aligned}
2 &= 1^2 + 1^2 = (1+i)(1-i) = -i(1+i)^2 && \text{so 2 is ramified}\\
5 &= 2^2 + 1^2 = (2+i)(2-i) && \text{5 is split}\\
13 &= 3^2 + 2^2 = (3+2i)(3-2i) && \text{13 is split}\\
17 &= 4^2 + 1^2 = (4+i)(4-i) && \text{17 is split}\\
29 &= 5^2 + 2^2 = (5+2i)(5-2i) && \text{29 is split}
\end{aligned}
$$

whereas 3, 7, 11, 19, 23 and 31 are inert in $\mathbb{Z}[i]$, since we cannot write these primes in the form x^2+y^2.

If we choose a different quadratic ring, then the pattern of which primes are inert, ramified and split will be quite different. For example, consider the case $d = -3$ so $\alpha = \frac{1+\sqrt{-3}}{2}$. The quadratic ring $\mathbb{Z}[\alpha]$ is norm-Euclidean, so it has unique factorization. By Equation (5.1), the norm is given by

$$\mathrm{N}(x + y\alpha) = x^2 + xy + y^2.$$

As the ring is complex, there are no elements of negative norm. If $\mathrm{N}(x + y\alpha) = p$, then $p = (x + y\alpha)(x + y\bar{\alpha})$. Since $\bar{\alpha} = 1 - \alpha$, this factorization is $(x + y\alpha)(x + y - y\alpha)$. The first few primes decompose in $\mathbb{Z}[\alpha]$ as follows:

$$\begin{array}{rll}
2 & & 2 \text{ is inert} \\
3 = -(\sqrt{-3})^2 & = -(2\alpha - 1)^2 & 3 \text{ is ramified} \\
5 & & 5 \text{ is inert} \\
7 = \mathrm{N}(2 + \alpha) & = (2 + \alpha)(3 - \alpha) & 7 \text{ is split} \\
11 & & 11 \text{ is inert} \\
13 = \mathrm{N}(3 + \alpha) & = (3 + \alpha)(4 - \alpha) & 13 \text{ is split.}
\end{array}$$

The next theorem predicts exactly which primes are inert, which are split and which are ramified.

Theorem 5.3 (The Decomposition Theorem). *Let $\mathbb{Z}[\alpha]$ be a quadratic ring with unique factorization and let p be an odd prime number.*

(1) *p is ramified if and only if p is a factor of d.*
(2) *p splits if and only if $\left(\frac{d}{p}\right) = 1$.*
(3) *p is inert if and only if $\left(\frac{d}{p}\right) = -1$.*

The prime number 2 splits if $d \equiv 1 \bmod 8$; it is inert if $d \equiv 5 \bmod 8$, and it is ramified in all other cases.

Proof. The theorem has quite a lot of special cases which are proved in roughly the same way to each other. Rather than looking at every special case, we'll assume for simplicity that $d \not\equiv 1 \bmod 4$, so the quadratic ring is $\mathbb{Z}[\sqrt{d}]$, and the norm is given by $\mathrm{N}(x + y\sqrt{d}) = x^2 - dy^2$. The other case is similar.

(1) Suppose first that p is a factor of d, either even or odd. We'll show that p is ramified. Let Q be an irreducible factor of p. To show that p is ramified, it's sufficient to show that Q^2 is a factor of p.
Note that Q is a factor of $d = (\sqrt{d})^2$. Since Q is irreducible, Q must be a factor of $\sqrt{d}$, and therefore Q^2 is a factor of d. Since d is square-free, the highest common factor of d and p^2 is p, so we can find integers h, k such that $p = hd + kp^2$. Since Q^2 is a common factor of d and p^2, it follows that Q^2 is a factor of p, so p is ramified.

(2) Next consider the prime $p = 2$ in the cases that d is odd. As we're assuming $d \not\equiv 1 \bmod 4$, we are trying to show that 2 is ramified. Let Q be an irreducible factor of 2 in $\mathbb{Z}[\alpha]$; again we must show that Q^2 is a factor of 2.
Note that $(1+\sqrt{d})^2 = 1+d+2\sqrt{d}$, which is a multiple of 2. This shows that Q is a factor of $(1+\sqrt{d})^2$, and hence of $1+\sqrt{d}$. Similarly, we know that Q is a factor of $1-\sqrt{d}$. Hence, Q^2 is a factor of $(1+\sqrt{d})(1-\sqrt{d}) = 1-d$. Since $\mathrm{hcf}(1-d, 4) = 2$, we can find integers h and k such that

$$4h + (1-d)k = 2.$$

Since Q^2 is a common factor of 4 and $1-d$, it follows that Q^2 is a factor of 2, so 2 is ramified.

(3) So far, we have shown that if p is a factor of $2d$, then p is ramified. Conversely, assume that p is ramified. Then there is an element $Q = x + y\sqrt{d}$, such that Q^2 is a multiple of p, but Q is not a multiple of p. Since $Q^2 = (x^2 + dy^2) + 2xy\sqrt{d}$, we know that

$$x^2 + dy^2 \equiv 0 \bmod p, \quad 2xy \equiv 0 \bmod p,$$

but x and y are not both multiples of p. In fact, y is not a multiple of p, since otherwise the first equation above would force x to also be a multiple of p. Since y is invertible modulo p, the second equation implies $2x \equiv 0 \bmod p$. Hence, by the first equation, we have $4dy^2 \equiv 0 \bmod p$. Since y is invertible, it follows that $4d \equiv 0 \bmod p$, so p is a factor of $2d$.

(4) From now on we shall assume that p is not a factor of $2d$, so p is either inert or split.
Suppose p splits, so there is an element $Q = x + y\sqrt{d}$ with norm $\pm p$. Then, we have

$$x^2 - dy^2 = \pm p.$$

Suppose p is a factor of y. Then we have $x^2 \equiv \pm p \bmod p^2$. This implies that p is a factor of x, and therefore $0 \equiv \pm p \bmod p^2$, which gives a contradiction. From this, we deduce that y is invertible modulo p, and therefore

$$\left(\frac{x}{y}\right)^2 \equiv d \bmod p.$$

In particular, d is a quadratic residue modulo p.

(5) Conversely suppose $x^2 \equiv d \bmod p$ and let $A = x + \sqrt{d}$. Clearly p is a factor of $x^2 - d = A\bar{A}$, but p is a factor of neither A nor $\bar{A}$. Hence, p is not irreducible so it splits. □

A criterion for uniqueness of factorization. One can use the Decomposition Theorem to check whether a quadratic ring has unique factorization or not. For example, let's consider the ring $\mathbb{Z}[\sqrt{10}]$. If this ring had unique factorization, then according to the decomposition theorem the primes 2 and 5 would be ramified in this ring, and every prime p for which $\left(\frac{10}{p}\right) = 1$ would split. This would imply that there are elements of norm $\pm p$ for all such primes p. Let's check whether this is the case. The norm of a general element $x + y\sqrt{10}$ is $x^2 - 10y^2$, so we consider the equation,

$$x^2 - 10y^2 = \pm 2. \tag{5.5}$$

If we reduce this modulo 5, then we see that $x^2 \equiv \pm 2 \bmod 5$. However, this congruence has no solutions since both 2 and -2 are quadratic non-residues modulo 5. Therefore, Equation (5.5) has no solutions in integers, and so there is no element of norm 2 or -2 in $\mathbb{Z}[\sqrt{10}]$. As a consequence of this, we see that 2 does not factorize in $\mathbb{Z}[\sqrt{10}]$, and so $\mathbb{Z}[\sqrt{10}]$ does not have unique factorization.

The method used above is particularly effective in the case of complex quadratic rings. This is because it's quite easy (for example) to see that there are no solutions in integers to $x^2 + 10y^2 = 5$, and so it follows immediately that $\mathbb{Z}[\sqrt{-10}]$ does not have unique factorization.

It turns out that one can also show that a ring *does* have unique factorization, just by checking the predictions of the Decomposition Theorem. To clarify this, note that the decomposition theorem predicts exactly which prime numbers factorize in $\mathbb{Z}[\alpha]$. Let

$$S_d = \left\{\text{primes } p : p|d \text{ or } \left(\frac{d}{p}\right) = 1 \text{ or } (p = 2 \text{ and } d \not\equiv 5 \bmod 8)\right\}.$$

The decomposition theorem tells us that if $\mathbb{Z}[\alpha]$ has unique factorization, then for every prime number $p \in S_d$, there is an element with norm $\pm p$. It turns out that the converse is true, and this gives us a way of proving that $\mathbb{Z}[\alpha]$ has unique factorization, even in cases where the ring is not norm-Euclidean. Of course, the set S_d is infinite so it is still a lot of work to check all the primes in S_d. But in fact one can say even more: one only needs to check finitely many of the primes in S_d. More precisely, define a positive real number M_d as follows:

$$M_d = \begin{cases} \sqrt{|d|} & \text{if } d > 0 \text{ and } d \equiv 1 \bmod 4, \\ 2\sqrt{|d|} & \text{if } d > 0 \text{ and } d \not\equiv 1 \bmod 4, \\ \frac{2}{\pi}\sqrt{|d|} & \text{if } d < 0 \text{ and } d \equiv 1 \bmod 4, \\ \frac{4}{\pi}\sqrt{|d|} & \text{if } d < 0 \text{ and } d \not\equiv 1 \bmod 4. \end{cases}$$

The number M_d is called the *Minkowski bound.*

Theorem 5.4. *Let $\mathbb{Z}[\alpha]$ be a quadratic ring and let M_d be the corresponding Minkowski bound. Then $\mathbb{Z}[\alpha]$ has unique factorization if and only if for every prime number $p \in S_d$ with $p < M_d$ there is an element of $\mathbb{Z}[\alpha]$ with norm $\pm p$.*

The most natural proof of this theorem uses the concept of ideals, which we have not covered in this book. One can prove the theorem easily using results from any book on algebraic number theory, for example [16] or [21].

The ring $\mathbb{Z}[\frac{1+\sqrt{-163}}{2}]$. As an example, we'll show that the quadratic ring $\mathbb{Z}[\frac{1+\sqrt{-163}}{2}]$ has unique factorization, even though this ring is very far from being norm-Euclidean. The Minkowski bound in this case is

$$M_{-163} = \frac{2\sqrt{163}}{\pi} \approx 25.534.$$

The primes less than this bound are

$$2, 3, 5, 7, 11, 13, 17, 19, 23.$$

Since $-163 \equiv 5 \bmod 8$, the prime 2 is not in S_{-163}. The number 163 is prime, so none of the primes listed above are factors of 163. Furthermore, we can check that $\left(\frac{-163}{p}\right) = -1$ for each odd prime in the list. This can be done by hand or on `sage` using the following code:

```
for p in prime_range(3,25):
    print(p,legendre_symbol(-163,p))
```

Therefore, there are no primes less than the Minkowski bound in S_{-163}. By Theorem 5.4, the ring has unique factorization.

Related to this example, note that we have

$$\mathrm{N}(x + \alpha) = x^2 + x + 41.$$

The polynomial $m(x) = x^2 + x + 41$ is rather famous because all of the numbers $m(0), m(1), \ldots, m(39)$ are prime. We'll now explain why this happens.

Corollary 5.2. *For $x = 0, 1, \ldots, 39$, the number $x^2 + x + 41$ is prime.*

Proof. Let x be an integer between 0 and 39. Suppose $x^2 + x + 41$ is composite, and let p be its smallest prime factor. Then we have $p^2 \leq x^2 + x + 41$, and this implies $p < 40$. If we let $\alpha = \frac{1+\sqrt{-163}}{2}$, then neither the number $x + \alpha$ nor $x + \bar{\alpha}$ is a multiple of p in $\mathbb{Z}[\alpha]$. However, their product is $m(x)$, which is a multiple of p. As $\mathbb{Z}[\alpha]$ has unique factorization, it follows that p is not irreducible in $\mathbb{Z}[\alpha]$, so there must be some element $r + s\alpha$ with norm p. In particular, we have $r^2 + rs + 41s^2 = p$. Completing the square, we see that $40s^2 \leq p < 40$, and therefore $s = 0$. This implies $r^2 = p$, which gives a contradiction since p is prime. □

Diophantine equations of the form $|\mathrm{N}(A)| = n$. We'll now answer the question: which integers can be written as $|\mathrm{N}(A)|$ for some $A \in \mathbb{Z}[\alpha]$. We shall restrict ourselves to the case that $\mathbb{Z}[\alpha]$ has unique factorization; other cases are more difficult and are described in the book [5].

Example. Let $d = -1$, so the quadratic ring is the ring $\mathbb{Z}[i]$ of Gaussian integers. Then $\mathrm{N}(x+iy) = x^2+y^2$, so the question becomes: Which integers have the form $x^2 + y^2$ with $x, y \in \mathbb{Z}$? Here are some examples:

$$\begin{aligned} 2 &= 1^2 + 1^2 \\ 4 &= 2^2 + 0^2 \\ 5 &= 2^2 + 1^2 \\ 8 &= 2^2 + 2^2 \\ 9 &= 3^2 + 0^2 \\ 10 &= 3^2 + 1^2. \end{aligned}$$

However, the numbers 3, 6 and 7 cannot be written as a sum of two squares.

The answer follows easily from the Decomposition Theorem. In general, an element A of $\mathbb{Z}[\alpha]$ is a product of irreducible elements. Hence, the norm of A is a product of norms of irreducible elements. If p is ramified or split, then there is an irreducible element of norm $\pm p$. If p is inert, then it is irreducible and has norm p^2. Hence, the power of every inert prime in $\mathrm{N}(A)$ is even.

Corollary 5.3. *Let $\mathbb{Z}[\alpha]$ be a quadratic ring with unique factorization. Let n be a positive integer. The following are equivalent:*

(1) *There is an $A \in \mathbb{Z}[\alpha]$ such that $|\mathrm{N}(A)| = n$.*
(2) *For each prime p dividing n, if p is inert in $\mathbb{Z}[\alpha]$, then $v_p(n)$ is even.*

Proof. Assume $|\mathrm{N}(A)| = n$ and factorize A into irreducible elements:

$$A = Q_1 \cdots Q_r.$$

Then $|\mathrm{N}(Q_i)|$ is either a prime number p_i, or p_i^2 if p_i is inert. Hence,

$$|\mathrm{N}(A)| = \prod_{\substack{\text{inert} \\ \text{primes } p_i}} p_i^2 \prod_{\substack{\text{split or ramified} \\ \text{primes } p_i}} p_i.$$

In particular, the powers of inert primes are even.

Conversely, suppose

$$n = \prod_{\substack{\text{inert} \\ \text{primes } p_i}} p_i^{2r_i} \prod_{\substack{\text{split or ramified} \\ \text{primes } p_i}} p_i^{r_i}.$$

Choose irreducible elements Q_i such that $|\mathrm{N}(Q_i)| = p_i$ or p_i^2. If we define $A = \prod Q_i^{r_i}$, then clearly $|\mathrm{N}(A)| = n$. □

Example. We'll write 585 as a sum of two squares in all possible ways. This is equivalent to finding all Gaussian integers of norm 585. We have

$$585 = 3^2 \cdot 5 \cdot 13.$$

The prime 3 is inert in $\mathbb{Z}[i]$. This means that 3 is irreducible and has norm 3^2. The primes 5 and 13 both split, with factors of norm 5 and 13.

$$5 = (2+i)(2-i), \quad 13 = (3+2i)(3-2i).$$

This means that every element of norm 5 is a unit multiple of $2+i$ or $2-i$, and every element of norm 13 is a unit multiple of $3+2i$ or $3-2i$. Therefore, the elements of norm 585 are unit multiples of one of the following elements:

$$\begin{aligned}
3(2+i)(3+2i) &= 12+21i, \\
3(2+i)(3-2i) &= 24-3i, \\
3(2-i)(3+2i) &= 24+3i, \\
3(2-i)(3-2i) &= 12-21i.
\end{aligned}$$

As $\mathbb{Z}[i]$ has four units, there are 16 solutions to the equation $x^2+y^2 = 585$. However, many of these solutions look very much alike. If we assume that x and y are both positive, and $x \leq y$, then we just have two solutions:

$$585 = 12^2 + 21^2 = 3^2 + 24^2.$$

The other solutions are obtained by swapping the variables x and y, or by changing their signs.

Example. Consider the equation $x^2+xy+y^2=150$. Note that $x^2+xy+y^2=\mathrm{N}(x+y\alpha)$ in the case $d=-3$ (the Eisenstein integers). The number 150 factorizes as $2\times 3\times 5^2$. By the decomposition theorem, 2 is inert, 3 is ramified and 5 is inert. Since $v_2(150)=1$, which is odd, the equation has no solutions in integers.

Example. We'll solve the equation $x^2+xy+y^2=84$. We begin with the factorization $84=2^2\times 3\times 7$. We have $\left(\frac{-3}{7}\right)=1$. By the Decomposition Theorem, 2 is inert, 3 is ramified and 7 is split. The specific factorizations of 3 and 7 are:

$$3=-\sqrt{-3}^2=-(2\alpha-1)^2 \quad \text{and} \quad 7=\mathrm{N}(2+\alpha)=(2+\alpha)(3-\alpha).$$

The Eisenstein integers of norm 84 are unit multiples of either $2(2\alpha-1)(2+\alpha)$ or $2(2\alpha-1)(3-\alpha)$. Using the fact that $\alpha^2+\alpha+1=0$, we find that these two numbers are $10\alpha-8$ and $2-10\alpha$. There are six units in the Eisenstein integers, so there are in total 12 elements of norm 84. For example,

$$84=\mathrm{N}(10\alpha-8)=10^2+10\cdot(-8)+(-8)^2.$$

Exercise 5.13. Let $\alpha=\frac{1+\sqrt{-19}}{2}$.

(1) Show using Theorem 5.4, that $\mathbb{Z}[\alpha]$ has unique factorization.
(2) Which of the following primes are (i) split in $\mathbb{Z}[\alpha]$, (ii) inert in $\mathbb{Z}[\alpha]$ and (iii) ramified in $\mathbb{Z}[\alpha]$?

$$2,\ 3,\ 5,\ 7,\ 11,\ 13,\ 17,\ 19,\ 23.$$

Factorize each split or ramified prime into irreducible elements.
(3) Which of the following integers may be written in the form $x^2+xy+5y^2$ with $x,y\in\mathbb{Z}$?

$$1540, \quad 399, \quad 210, \quad 115.$$

Exercise 5.14. Which of the following numbers may be written in the form x^2+xy-y^2 with $x,y\in\mathbb{Z}$?

$$-1, \quad 20, \quad -20, \quad 4141, \quad 35055.$$

Exercise 5.15.

(1) Show that there is no element of $\mathbb{Z}[\sqrt{3}]$ with norm -1.
(2) Let p be a prime number. Show that if there is an element in $\mathbb{Z}[\sqrt{3}]$ with norm p, then there is no element of norm $-p$.

(3) Let p be a prime number. Show that there is a solution to $x^2 - 3y^2 = p$ if and only if $p \equiv 1 \bmod 12$.
(4) Which of the following integers may be written in the form $x^2 - 3y^2$ with $x, y \in \mathbb{Z}$?

$$66, \quad -66, \quad 1573, \quad -1573.$$

Find a solution in each case where one exists.

Exercise 5.16. Let $\mathbb{Z}[\alpha]$ be a quadratic ring (with or without unique factorization). Prove that every non-zero element of $\mathbb{Z}[\alpha]$ is either a unit or a product of irreducible elements. Hence, prove that there are infinitely many irreducible elements in $\mathbb{Z}[\alpha]$.

Exercise 5.17. Using Theorem 5.4, determine whether $\mathbb{Z}[\alpha]$ has unique factorization in the cases $d = -63, 33, 34$.

Exercise 5.18. Let p and q be distinct primes with $p \equiv 1 \bmod 4$. Show that $\mathbb{Z}[\alpha_{pq}]$ does not have unique factorization. You may assume there is a prime number l such that $\left(\frac{p}{l}\right) = \left(\frac{q}{l}\right) = -1$ (this follows from Dirichlet's primes in arithmetic progressions theorem).

Exercise 5.19. Let $\mathbb{Z}[\alpha]$ be a quadratic ring with unique factorization, and assume that d is odd and positive. Show using Dirichlet's primes in arithmetic progressions theorem that there are infinitely many prime numbers which are inert in $\mathbb{Z}[\alpha]$. Similarly, show that there are infinitely many primes which split in $\mathbb{Z}[\alpha]$. (It's not necessary to assume that d is odd and positive; this just makes the question shorter by excluding a lot of special cases.)

5.5 CONTINUED FRACTIONS

We showed earlier that the complex quadratic rings do not have many units; apart from two special cases, their units are only 1 and -1. Our next aim is to determine the units in the real quadratic rings. Here, the situation is rather different. We'll show that every real quadratic ring has infinitely many units. For example, to find units in the ring $\mathbb{Z}[\sqrt{6}]$, we need to solve the equation:

$$x^2 - 6y^2 = \pm 1. \tag{5.6}$$

If x and y are large positive integers which solve this equation, then we see that $\frac{x}{y}$ is a very good approximation to $\sqrt{d}$. In this section, we'll discuss continued fractions, which are sequences of rational approximations to a real number. In the next section, we'll use the continued fraction method to solve equations such as Equation (5.6), and hence find the units in real quadratic rings.

Finite continued fractions. By a *finite continued fraction*, we shall mean an expression of the form

$$a_0 + \cfrac{1}{a_1 + \cfrac{1}{a_2 + \cfrac{1}{\ddots + \cfrac{1}{a_n}}}}, \qquad \text{where } a_0, a_1, \ldots, a_n \in \mathbb{Z} \text{ and } a_1, \ldots, a_n > 0.$$

Such expressions are difficult to write, so we'll use the following notation to mean the same thing:

$$[a_0, a_1, \ldots, a_n].$$

For example, we have

$$[1,2,3,2] = 1 + \cfrac{1}{2 + \cfrac{1}{3 + \cfrac{1}{2}}} = 1 + \cfrac{1}{2 + \cfrac{2}{7}} = 1 + \frac{7}{16} = \frac{23}{16}.$$

More generally, for real numbers $\alpha > 0$, we'll sometimes use the notation

$$[a_0, a_1, \ldots, a_n, \alpha] = a_0 + \cfrac{1}{a_1 + \cfrac{1}{a_2 + \cfrac{1}{\ddots + \cfrac{1}{a_n + \cfrac{1}{\alpha}}}}}.$$

This has the following useful consequence for any $r < n$:

$$[a_0, \ldots, a_n] = [a_0, \ldots, a_r, [a_{r+1}, \ldots, a_n]].$$

Obviously, every finite continued fraction is a rational number. It turns out that the converse is also true, and we can write every rational number as a finite continued fraction. To illustrate how this is done, we'll write $\frac{89}{35}$ as a finite continued fraction. Our method is to go through Euclid's algorithm with 89 and 35; next to each line $n = qm + r$, we also write the equivalent $\frac{n}{m} = q + \frac{1}{(\frac{m}{r})}$.

$$89 = \underline{\mathbf{2}} \times 35 + 19 \qquad \frac{89}{35} = \underline{\mathbf{2}} + \frac{1}{(\frac{35}{19})}$$

$$35 = \underline{\mathbf{1}} \times 19 + 16 \qquad \frac{35}{19} = \underline{\mathbf{1}} + \frac{1}{(\frac{19}{16})}$$

$$19 = \underline{\mathbf{1}} \times 16 + 3 \qquad \frac{19}{16} = \underline{\mathbf{1}} + \frac{1}{(\frac{16}{3})}$$

$$16 = \underline{\mathbf{5}} \times 3 + 1 \qquad \frac{16}{3} = \underline{\mathbf{5}} + \frac{1}{3}$$

$$3 = \underline{\mathbf{3}} \times 1 + 0.$$

Each equation in the right-hand column may be substituted into the previous one as follows:

$$\frac{89}{35} = \underline{\mathbf{2}} + \cfrac{1}{\underline{\mathbf{1}} + \cfrac{1}{\underline{\mathbf{1}} + \cfrac{1}{\underline{\mathbf{5}} + \cfrac{1}{\underline{\mathbf{3}}}}}}.$$

This shows that $\frac{89}{35} = [2, 1, 1, 5, 3]$. The numbers $2, 1, 1, 5, 3$ in the continued fraction are exactly the quotients in Euclid's algorithm; these numbers are sometimes called the *partial quotients* of the continued fraction.

Infinite continued fractions. Suppose now that we have an infinite sequence of integers a_n, such that a_n is positive for all $n > 0$. We define a sequence of rational numbers x_n by

$$x_n = [a_0, \ldots, a_n].$$

For example, the first few x_n are

$$\begin{aligned} x_0 &= a_0, \\ x_1 &= a_0 + \frac{1}{a_1} = \frac{a_0 a_1 + 1}{a_1}, \\ x_2 &= a_0 + \frac{1}{a_1 + \frac{1}{a_2}} = a_0 + \frac{a_2}{a_1 a_2 + 1} = \frac{(a_0 a_1 + 1)a_2 + a_0}{a_1 a_2 + 1}. \end{aligned}$$

The rational numbers x_n are called *convergents*. We'll show that the convergents x_n converge to a limit in $\mathbb{R}$. It will be useful to have formulae for the numerator and denominator of x_n. We'll define sequences of integers h_n and k_n recursively by

$$\begin{aligned} h_0 &= a_0, & h_1 &= a_1 a_0 + 1, & h_n &= a_n h_{n-1} + h_{n-2}, \\ k_0 &= 1, & k_1 &= a_1, & k_n &= a_n k_{n-1} + k_{n-2} \qquad (n \geq 2). \end{aligned}$$

Proposition 5.3. *With the notation above, we have* $[a_0, a_1, \ldots, a_n] = \frac{h_n}{k_n}$, *and more generally*

$$[a_0, a_1, \ldots, a_n, \alpha] = \frac{\alpha h_n + h_{n-1}}{\alpha k_n + k_{n-1}}.$$

Proof. We can deduce the first formula from the second formula by substituting $\alpha = a_{n+1}$. We'll prove the second formula by induction on n. The formula is true for $n = 1$ by the following calculation:

$$[a_0, a_1, \alpha] = a_0 + \frac{1}{a_1 + \frac{1}{\alpha}} = \frac{(a_0 a_1 + 1)\alpha + a_0}{a_1 \alpha + 1}.$$

Assume the formula is correct for n. We have

$$[a_0, a_1, \ldots, a_n, a_{n+1}, \alpha] = \left[a_0, a_1, \ldots, a_{n+1} + \frac{1}{\alpha}\right].$$

The formula on the right-hand side is a continued fraction of length n, and so by the inductive hypothesis it is equal to

$$\frac{(a_{n+1} + \frac{1}{\alpha})h_n + h_{n-1}}{(a_{n+1} + \frac{1}{\alpha})k_n + k_{n-1}}.$$

Multiplying the numerator and denominator by α, we get

$$[a_0, \ldots, a_{n+1}, \alpha] = \frac{\alpha(a_{n+1} h_n + h_{n-1}) + h_{n-1}}{\alpha(a_{n+1} k_n + k_{n-1}) + k_{n-1}}.$$

By the recursive formula for h_{n+1} and k_{n+1}, we have $[a_0, \ldots, a_{n+1}, \alpha] = \frac{\alpha h_{n+1} + h_n}{\alpha k_{n+1} + k_n}$. □

We can see from the definition, that the denominators k_n form a strictly increasing sequence of positive integers. The signs of the numerators h_n depend on the sign of a_0. We'll see in the next result that the fractions $x_n = \frac{h_n}{k_n}$ do not cancel.

Lemma 5.5. *The integers h_n and k_n are coprime. Furthermore, we have*

$$h_{n+1} k_n - k_{n+1} h_n = (-1)^n.$$

Proof. Note that the formula implies coprimality, so we'll just prove the formula. We can prove this by induction on n. It's trivial to check the formula in the cases $n = 0$ and $n = 1$. Assume the formula for $n - 1$. By the recursive formulae, we have

$$\begin{aligned} h_{n+1}k_n - k_{n+1}h_n &= (a_{n+1}h_n + h_{n-1})k_n - (a_{n+1}k_n + k_{n-1})h_n \\ &= -(h_n k_{n-1} - k_n h_{n-1}). \end{aligned}$$

The result follows from the inductive hypothesis. □

Theorem 5.5. *Let $a_0, a_1, \ldots$ be any sequence of integers with $a_n > 0$ for $n > 0$. Then the sequence $x_n = [a_0, \ldots, a_n]$ converges to a limit $\alpha \in \mathbb{R}$. The limit α is between x_n and x_{n+1}, and we have $|\alpha - x_n| < \frac{1}{k_n^2}$.*

To prove this, we'll use the following result from real analysis:

Proposition 5.4 (Alternating series test). *Let y_n be a decreasing sequence of positive real numbers such that $y_n \to 0$ as $n \to \infty$. Then, the series $\alpha = \sum_{n=1}^{\infty}(-1)^n y_n$ converges. For any n, the limit α is between the n-th and $n+1$-st partial sums.*

Proof of Theorem 5.5. Note that by Lemma 5.5, we have

$$x_{n+1} - x_n = \frac{h_{n+1}}{k_{n+1}} - \frac{h_n}{k_n} = \frac{(-1)^n}{k_n k_{n+1}}.$$

Using this, we have

$$\begin{aligned} x_n &= x_0 + (x_1 - x_0) + \cdots + (x_n - x_{n-1}) \\ &= a_0 + \frac{1}{k_0 k_1} - \frac{1}{k_1 k_2} + \frac{1}{k_2 k_3} - \cdots + \frac{(-1)^{n-1}}{k_{n-1}k_n}. \end{aligned}$$

Therefore,

$$\lim_{n\to\infty} x_n = a_0 + \sum_{n=0}^{\infty} \frac{(-1)^n}{k_n k_{n+1}},$$

and this converges by the alternating series test. Also by that test, we know that the limit is between the n-th and $n+1$-st partial sums, i.e. between x_n and x_{n+1}. This implies

$$|\alpha - x_n| < |x_{n+1} - x_n| = \frac{1}{k_n k_{n+1}} < \frac{1}{k_n^2}.$$ □

We shall use the notation $[a_0, a_1, \ldots]$ for the limit of the sequence of convergents $\frac{h_n}{k_n}$. Such limits are called *infinite continued fractions*. We can sometimes calculate the values of these limits. For example, consider the real number

$$\alpha = [1,2,1,2,1,2,1,2,\ldots] = 1 + \cfrac{1}{2 + \cfrac{1}{1 + \ddots}}.$$

It's clear that $\alpha > 0$; indeed all the convergents are bigger than 1. Therefore, the function $x \mapsto [1,2,x]$ is continuous at $x = \alpha$ (see Exercise 5.22). This implies $\lim_{n\to\infty}[1,2,x_n] = [1,2,\lim_{n\to\infty} x_n]$, so we have

$$\alpha = [1,2,\alpha] = 1 + \frac{1}{2+\frac{1}{\alpha}}.$$

Simplifying the continued fraction, we get $\alpha = \frac{3\alpha+1}{2\alpha+1}$. Therefore,

$$2\alpha^2 - 2\alpha - 1 = 0.$$

The quadratic equation has two solutions $\alpha = \frac{1\pm\sqrt{3}}{2}$. Since α is positive, we have $\alpha = \frac{1+\sqrt{3}}{2}$.

Irrational numbers as continued fractions. We've shown that every rational number may be written as a finite continued fraction. We'll next show that every irrational real number α may be written as the limit of an infinite continued fraction. To find the continued fraction expansion of α, we define two sequences $\alpha_n \in \mathbb{R}$ and $a_n \in \mathbb{Z}$ recursively as follows:

$$\alpha_0 = \alpha, \qquad a_n = \lfloor \alpha_n \rfloor, \qquad \alpha_{n+1} = \frac{1}{\alpha_n - a_n}.$$

Here, the notation $\lfloor \alpha_n \rfloor$ means the *floor* of α_n, i.e. the greatest integer which is less than or equal to α_n. Note that a_n will not actually be equal to α_n because α_n is irrational. This means that we will never need to divide by zero in this process. Also, since $\alpha_n - a_n$ is between 0 and 1, we know that $\alpha_{n+1} > 1$. Therefore, the integers a_n (apart from possibly a_0) must all be positive.

Proposition 5.5. *With the notation defined above,* $\alpha = [a_0, a_1, a_2, \ldots]$.

Proof. Note that we have

$$\alpha = \alpha_0 = a_0 + \frac{1}{\alpha_1} = a_0 + \frac{1}{a_1 + \frac{1}{\alpha_2}} = \ldots = [a_0, a_1, \ldots, a_n, \alpha_{n+1}].$$

By Proposition 5.3, we have $\alpha = \frac{\alpha_{n+1}h_n + h_{n-1}}{\alpha_{n+1}k_n + k_{n-1}}$. This implies

$$\left|\alpha - \frac{h_n}{k_n}\right| = \left|\frac{(\alpha_{n+1}h_n + h_{n-1})k_n - (\alpha_{n+1}k_n + k_{n-1})h_n}{(\alpha_{n+1}k_n + k_{n-1})k_n}\right|.$$

By Lemma 5.5, the numerator on the right-hand side is ± 1, and therefore

$$\left|\alpha - \frac{h_n}{k_n}\right| = \frac{1}{(\alpha_{n+1}k_n + k_{n-1})k_n}.$$

The right-hand side tends to zero, because k_n is an increasing sequence of positive integers. Therefore, $\frac{h_n}{k_n}$ converges to α. □

As an example, we'll calculate the continued fraction expansion of $\sqrt{2}$. Following the method described above, we define sequences α_n and a_n as follows:

$$\alpha_0 = \sqrt{2}, \quad a_n = \lfloor \alpha_n \rfloor, \quad \alpha_{n+1} = \frac{1}{\alpha_n - a_n}.$$

We have $a_0 = \lfloor \sqrt{2} \rfloor = 1$, and so

$$\alpha_1 = \frac{1}{\sqrt{2} - 1}.$$

To calculate the next integer $a_1 = \lfloor \alpha_1 \rfloor$, it will be useful to simplify our expression for α_1. To do this, we multiply the numerator and denominator by $\sqrt{2} + 1$, in order to cancel the square root in the denominator:

$$\alpha_1 = \frac{\sqrt{2} + 1}{(\sqrt{2} - 1)(\sqrt{2} + 1)} = \sqrt{2} + 1.$$

From this, we can see that $a_1 = \lfloor \sqrt{2} + 1 \rfloor = 2$. Next, note that

$$\alpha_2 = \frac{1}{(\sqrt{2} + 1) - 2} = \frac{1}{\sqrt{2} - 1}.$$

At this point, we notice that $\alpha_2 = \alpha_1$. From the recursive formula $\alpha_{n+1} = \frac{1}{\alpha_n - \lfloor \alpha_n \rfloor}$, it follows that $\alpha_1 = \alpha_2 = \alpha_3 = \cdots = \sqrt{2} + 1$. This implies $a_n = 2$ for all $n \geq 1$, so we have

$$\sqrt{2} = [1, 2, 2, 2, 2, \ldots].$$

As a result of this, we get a sequence of rational approximations to $\sqrt{2}$:

$$[1, 2] = \frac{3}{2} = 1.5$$

$$[1, 2, 2] = \frac{7}{5} = 1.4$$

$$[1,2,2,2] = \frac{17}{12} \approx 1.416666667$$
$$[1,2,2,2,2] = \frac{41}{29} \approx 1.413793103$$
$$[1,2,2,2,2,2] = \frac{99}{70} \approx 1.414285714.$$

Note that by Theorem 5.5, the accuracy of the approximation $\frac{h_n}{k_n}$ better than $\frac{1}{k_n^2}$, so, for example, we have

$$\left|\sqrt{2} - \frac{99}{70}\right| < \frac{1}{4900}.$$

In fact, there is nearly a converse to Theorem 5.5; if we have an especially accurate rational approximation to a real number α, then we can deduce that it is one of the convergents.

Theorem 5.6. *Let α be an irrational real number and $\frac{h}{k}$ a rational number satisfying*

$$\left|\alpha - \frac{h}{k}\right| < \frac{1}{2k^2}.$$

Then $\frac{h}{k}$ is one of the convergents in the continued fraction expansion of α.

The theorem requires the following rather tricky lemma in its proof.

Lemma 5.6. *Let α be an irrational real number and let $\frac{h_n}{k_n}$ be the n-th convergent in its continued fraction expansion. If $\frac{h}{k}$ is a rational number with $1 \leq k < k_{n+1}$ and $\frac{h}{k} \neq \frac{h_n}{k_n}$, then*

$$|k\alpha - h| > |k_n\alpha - h_n|.$$

Proof. Consider the following system of simultaneous equations:

$$h_n x + h_{n+1} y = h,$$
$$k_n x + k_{n+1} y = k.$$

By Lemma 5.5, the matrix $\begin{pmatrix} h_n & h_{n+1} \\ k_n & k_{n+1} \end{pmatrix}$ has determinant ± 1. Therefore, these equations have a unique solution $(x, y) \in \mathbb{Z}^2$. Note that neither x nor y is zero, since otherwise $\frac{h}{k}$ would be equal to $\frac{h_{n+1}}{k_{n+1}}$ or $\frac{h_n}{k_n}$. Furthermore, it follows from the second of the two simultaneous equations together with the fact that $0 < k < k_{n+1}$, that x and y have opposite signs.

By Theorem 5.5, the limit α is between $\frac{h_n}{k_n}$ and $\frac{h_{n+1}}{k_{n+1}}$. Therefore, $\alpha - \frac{h_n}{k_n}$ and $\alpha - \frac{h_{n+1}}{k_{n+1}}$ also have opposite signs. It follows that $x(k_n\alpha - h_n)$ and $y(k_{n+1}\alpha - h_{n+1})$ have the same sign. We now calculate:

$$\begin{aligned} |k\alpha - h| &= |(k_n x + k_{n+1} y)\alpha - (h_n x + h_{n+1} y)| \\ &= |x(k_n\alpha - h_n) + y(k_{n+1}\alpha - h_{n+1})| \\ &= |x| \cdot |k_n\alpha - h_n| + |y| \cdot |k_{n+1}\alpha - h_{n+1}| \\ &> |k_n\alpha - h_n|. \end{aligned}$$

(In the next to last line, we used the fact that both terms have the same sign.) □

Proof of Theorem 5.6. We'll assume that $\frac{h}{k}$ is not a convergent and derive a contradiction. Since the denominators k_n form an increasing sequence of natural numbers, there is an n, such that $k_n \leq k < k_{n+1}$. By Lemma 5.6, we have

$$|k_n\alpha - h_n| < |k\alpha - h| = k\left|\alpha - \frac{h}{k}\right| < \frac{1}{2k}.$$

Dividing by k_n, we get

$$\left|\alpha - \frac{h_n}{k_n}\right| < \frac{1}{2kk_n}.$$

Since $\frac{h}{k} \neq \frac{h_n}{k_n}$, it follows that

$$\left|\frac{h}{k} - \frac{h_n}{k_n}\right| \geq \frac{1}{kk_n}.$$

Hence, by the triangle inequality, we have

$$\frac{1}{kk_n} \leq \left|\frac{h}{k} - \frac{h_n}{k_n}\right| \leq \left|\frac{h}{k} - \alpha\right| + \left|\alpha - \frac{h_n}{k_n}\right| < \frac{1}{2k^2} + \frac{1}{2kk_n} \leq \frac{1}{kk_n}.$$

This is a contradiction. □

Exercise 5.20. Write the following rational numbers as finite continued fractions:

$$\frac{17}{45}, \frac{103}{456}, \frac{29}{135}.$$

This question can be answered in `sage` with the command

```
continued_fraction(17/45)
```

Exercise 5.21. The *Fibonacci sequence* is defined recursively by the formulae:

$$F_1 = F_2 = 1, \quad F_{n+2} = F_n + F_{n+1} \quad \text{for } n \geq 1.$$

Show that $\frac{F_{n+1}}{F_n} = [1, 1, \ldots, 1]$, where the number of 1's in the continued fraction is n. Hence, show that $\lim_{n\to\infty} \frac{F_{n+1}}{F_n} = \frac{1+\sqrt{5}}{2}$. (This limit is often called the *golden ratio.*)

Exercise 5.22. Show that the function $x \mapsto [a_0, a_1, \ldots, a_n, x]$ is continuous at all positive real numbers x.

Exercise 5.23. Calculate the following continued fractions:

$$[2, 2, 2, 2, 2, 2, \ldots], \quad [2, 3, 2, 3, 2, 3, 2, 3, \ldots], \quad [1, 2, 3, 5, 3, 5, 3, 5, \ldots].$$

Exercise 5.24. Let α be a real number, whose continued fraction is eventually periodic, i.e. $\alpha = [a_0, \ldots, a_r, \overline{a_{r+1}, \ldots, a_{r+s}}]$. Show that α is a root of a quadratic equation with rational coefficients.

Exercise 5.25. Let $\mathbb{Z}[\alpha_d]$ be a norm-Euclidean quadratic ring. Show that for every $Z \in \mathbb{Q}(\sqrt{d})$, there is a finite sequence of elements $a_0, \ldots, a_n \in \mathbb{Z}[\alpha]$ such that $Z = [a_0, \ldots, a_n]$.

5.6 PELL'S EQUATION

Let $d \geq 2$ be a square-free integer. *Pell's equation* is the Diophantine equation

$$x^2 - dy^2 = 1.$$

In this section, we'll see how to find all solutions in integers to this equation. There are obvious solutions $x = \pm 1$, $y = 0$. We'll call these the *trivial solutions.* However, it turns out that these equations all have non-trivial solutions as well. For example, when $d = 2$, we have the non-trivial solution $x = 3$, $y = 2$. For other values of d, the non-trivial solutions are less obvious. For example, when $d = 151$, the smallest non-trivial solution is $x = 1728148040$, $y = 140634693$.

There is another way of thinking about Pell's equation. If we set $A = x + y\sqrt{d}$, then Pell's equation says $\mathrm{N}(A) = 1$, and so integer solutions to Pell's equation correspond to units in quadratic rings. The trivial solutions $x = \pm 1$, $y = 0$ correspond to the obvious units 1 and -1 in $\mathbb{Z}[\sqrt{d}]$. By showing that Pell's equation has non-trivial solutions, we will be showing that real quadratic rings have more units than just 1 and -1. For example, since $3^2 - 2 \cdot 2^2 = 1$, we know that $\mathrm{N}(3 + 2 \cdot \sqrt{2}) = 1$. We can produce more units in $\mathbb{Z}[\sqrt{2}]$ with norm 1 by taking powers of $3 + 2\sqrt{2}$, for example

$$(3 + 2\sqrt{2})^2 = 17 + 12\sqrt{2},$$

$$(3 + 2\sqrt{2})^3 = 99 + 70\sqrt{2}.$$

Therefore, $(17, 12)$ and $(99, 70)$ are also solutions to $x^2 - 2y^2 = 1$, and we can produce infinitely many solutions by this process.

The fundamental solution. By the *fundamental solution* to Pell's equation, we shall mean the solution to $x^2 - dy^2 = 1$ with smallest positive x and y. It does not matter whether we minimize the variable x or the variable y, since $x = \sqrt{dy^2 + 1}$, which increases as y increases.

If we have a solution (x, y) to Pell's equation, then obviously there are three other solutions $(x, -y)$, $(-x, y)$ and $(-x, -y)$. We can recover the signs of the integers x and y if we know roughly the size of the real number $A = x + y\sqrt{d}$. If we assume that x and y are positive integers, then clearly $A = x + y\sqrt{d}$ is bigger than 1. Since $\mathrm{N}(A) = 1$, we know that $\bar{A} = x - y\sqrt{d} = \frac{1}{A}$. This implies that $x - y\sqrt{d}$ is between 0 and 1. Similarly $-x + y\sqrt{d}$ is between -1 and 0, and $-x - y\sqrt{d} < -1$. This tells us that if (x, y) is the fundamental solution to Pell's equation, then $A = x + y\sqrt{d}$ has norm 1 and A is the smallest element of $\mathbb{Z}[\sqrt{d}]$ with norm 1, such that $A > 1$.

Proposition 5.6. *Let (x, y) be the fundamental solution to Pell's equation. Then every solution has the form $(\pm x_n, \pm y_n)$, where*

$$x_n + y_n\sqrt{d} = (x + y\sqrt{d})^n.$$

Proof. Let $A = x + y\sqrt{d}$ and let $B = u + v\sqrt{d}$, where u, v is another solution to Pell's equation in positive integers. In particular, $B > 1$. Since $A > 1$, the sequence A^n is increasing and tends to infinity. Hence, there is an n such that $A^n \leq B < A^{n+1}$. This implies $1 \leq A^{-n}B < A$. Since $A^{-n}B$ is a unit with norm 1, it follows that $A^{-n}B = 1$, so $B = A^n$. □

Theorem 5.7. *For every square-free integer $d \geq 2$, Pell's equation has infinitely many solutions. Equivalently, every real quadratic ring $\mathbb{Z}[\alpha]$ has infinitely many units.*

Proof. The process described above produces infinitely many solutions to Pell's equation, given just one non-trivial solution. It's therefore sufficient to show that there is a non-trivial solution. Recall that for any convergent $\frac{h}{k}$ to $\sqrt{d}$, we have $|\frac{h}{k} - \sqrt{d}| < \frac{1}{k^2}$. Multiplying by $k(h + k\sqrt{d})$, we deduce:

$$|h^2 - dk^2| < \frac{h + k\sqrt{d}}{k} < 2\sqrt{d} + 1.$$

In particular, the integers $h^2 - dk^2$ are bounded. As there are infinitely many convergents $\frac{h}{k}$ but only finitely many possible values of $h^2 - dk^2$, there must be some integer n such that the equation,

$$h^2 - dk^2 = n, \tag{5.7}$$

has infinitely many integer solutions. Since there are only finitely many possibilities for h and k modulo n, we can find two distinct solutions (h, k) and (h', k') to Equation (5.7), such that $h \equiv h' \bmod n$ and $k \equiv k' \bmod n$. Now let

$$x + y\sqrt{d} = \frac{h + k\sqrt{d}}{h' + k'\sqrt{d}}.$$

Since $h+k\sqrt{d}$ and $h'+k'\sqrt{d}$ both have norm n, it follows that $\mathrm{N}(x+y\sqrt{d}) = 1$. Multiplying the numerator and denominator by $h' - k'\sqrt{d}$, we get

$$x + y\sqrt{d} = \frac{(h + k\sqrt{d})(h' - k'\sqrt{d})}{n} = \frac{hh' - dkk'}{n} + \frac{kh' - hk'}{n}\sqrt{d}.$$

By our congruence conditions on h, h', k and k', it follows that x and y are integers. Therefore, $x + y\sqrt{d}$ is an element of $\mathbb{Z}[\sqrt{d}]$ with norm 1. Equivalently, (x, y) is a non-trivial solution to Pell's equation. In the case that $d = 1 \bmod 5$, the ring $\mathbb{Z}[\sqrt{d}]$ is contained in $\mathbb{Z}[\alpha]$, so $\mathbb{Z}[\alpha]$ has infinitely many units in this case as well. □

Theorem 5.7 tells us that there is a fundamental solution to Pell's equation. Furthermore, Proposition 5.6 tells us how to write all solutions in terms of the fundamental solution. In a sense, this solves the problem completely; we can find the fundamental solution by checking the numbers $dy^2 + 1$ for $y = 1, 2, 3, \ldots$ until we come to the first square x^2 in this list. It follows that (x, y) is the fundamental solution. However, it is quite possible that the fundamental solution is large, so the method just described would take a long time. Instead, we can use the following result.

Proposition 5.7. *Suppose $x^2 - dy^2 = \pm 1$ with x and y positive integers. Then $\frac{x}{y}$ is a convergent of the continued fraction expansion of $\sqrt{d}$.*

Proof. We have $(x+y\sqrt{d})(x-y\sqrt{d}) = \pm 1$, and therefore $\left|\frac{x}{y} - \sqrt{d}\right| = \frac{1}{y(x+y\sqrt{d})}$. Since $x^2 - dy^2 = \pm 1$, it follows that $x > y$, and therefore $\left|\frac{x}{y} - \sqrt{d}\right| < \frac{1}{2y^2}$. Theorem 5.6 implies that $\frac{x}{y}$ is a convergent to $\sqrt{d}$. □

The proposition gives us a much faster way of searching for the fundamental solutions. Instead of listing all of the numbers dy^2+1 until we come to a square, we can calculate the continued fraction expansion of $\sqrt{d}$ and then list the convergents $\frac{h_n}{k_n}$ for $n = 0, 1, 2, \ldots$. The first convergent satisfying $h_n^2 - dk_n^2 = 1$ must be the fundamental solutions.

Example. Let $d = 7$, so we consider the equation

$$x^2 - 7y^2 = 1.$$

We'll calculate the continued fraction expansion of $\sqrt{7}$:

$$a_0 = \lfloor \sqrt{7} \rfloor = 2,$$

$$\alpha_1 = \frac{1}{\sqrt{7}-2} = \frac{\sqrt{7}+2}{(\sqrt{7}+2)(\sqrt{7}-2)} = \frac{\sqrt{7}+2}{3},$$

$$a_1 = \lfloor \alpha_1 \rfloor = 1,$$

$$\alpha_2 = \frac{1}{\frac{\sqrt{7}+2}{3} - 1} = \frac{3}{\sqrt{7}-1} = \frac{3(\sqrt{7}+1)}{(\sqrt{7}+1)(\sqrt{7}-1)} = \frac{3(\sqrt{7}+1)}{6} = \frac{\sqrt{7}+1}{2},$$

$$a_2 = \lfloor \alpha_2 \rfloor = 1,$$

$$\alpha_3 = \frac{1}{\frac{\sqrt{7}+1}{2} - 1} = \frac{2}{\sqrt{7}-1} = \frac{2(\sqrt{7}+1)}{(\sqrt{7}+1)(\sqrt{7}-1)} = \frac{2(\sqrt{7}+1)}{6} = \frac{\sqrt{7}+1}{3},$$

$$a_3 = \lfloor \alpha_3 \rfloor = 1,$$

$$\alpha_4 = \frac{1}{\frac{\sqrt{7}+1}{3} - 1} = \frac{3}{\sqrt{7}-2} = \frac{3(\sqrt{7}+2)}{(\sqrt{7}+2)(\sqrt{7}-2)} = \frac{3(\sqrt{7}+2)}{3} = \sqrt{7}+2,$$

$$a_4 = \lfloor \alpha_4 \rfloor = 4,$$

$$\alpha_5 = \frac{1}{\sqrt{7}-2} = \alpha_1.$$

At this point, we see that $\alpha_5 = \alpha_1$ and therefore $a_5 = a_1$. Furthermore, since $\alpha_{n+1} = \frac{1}{\alpha_n - \lfloor a_n \rfloor}$ it follows $\alpha_6 = \alpha_2$, $\alpha_7 = \alpha_3$, etc. Therefore, the partial quotients a_n repeat after this point, and we have

$$\sqrt{7} = [2, 1, 1, 1, 4, 1, 1, 1, 4, 1, 1, 1, 4, 1, 1, 1, 4, 1, 1, 1, \ldots].$$

We'll calculate the first few convergents $\frac{h}{k}$, together with the values of $h^2 - 7k^2$:

$$\frac{h_0}{k_0} = [2] = \frac{2}{1} \qquad 2^2 - 7 \cdot 1^2 = -3,$$

$$\frac{h_1}{k_1} = [2, 1] = 1 + \frac{1}{1} = \frac{3}{1} \qquad 3^2 - 7 \cdot 1^2 = 2,$$

$$\frac{h_2}{k_2} = [2, 1, 1] = 2 + \frac{1}{1 + \frac{1}{1}} = \frac{5}{2} \qquad 5^2 - 7 \cdot 2^2 = -3,$$

$$\frac{h_3}{k_3} = [2, 1, 1, 1] = 2 + \frac{1}{1 + \frac{1}{1+\frac{1}{1}}} = \frac{8}{3} \qquad 8^2 - 7 \cdot 3^2 = 1.$$

Hence, the fundamental solution to Pell's equation is $(8, 3)$.

The method described above is fairly practical for solving Pell's equation. However, it can be improved: There is a way of seeing immediately which convergents in the continued fraction expansion correspond to solutions to Pell's equation.

Theorem 5.8. *Let $\frac{h_n}{k_n} = [a_0, \ldots, a_n]$ be a convergent to $\sqrt{d}$. Then, $h_n + k_n\sqrt{d}$ is a unit in $\mathbb{Z}[\sqrt{d}]$ if and only if*

$$\sqrt{d} = [a_0, \overline{a_1, \ldots, a_n, 2a_0}].$$

If this is the case, then $\mathrm{N}(h_n + k_n\sqrt{d}) = (-1)^{n-1}$.

The notation $[a_0, \overline{a_1, \ldots, a_n, 2a_0}]$ means that the partial quotients $a_1, \ldots, a_n, 2a_0$ repeat forever. For example, we saw above that $\sqrt{7} = [2, \overline{1, 1, 1, 4}]$.

Proof. Suppose $h_n^2 - d \cdot k_n^2 = \pm 1$. We'll first work out how the sign depends on n. We saw in Theorem 5.5 that $\sqrt{d}$ lies between any two successive convergents $\frac{h_n}{k_n}$ and $\frac{h_{n+1}}{k_{n+1}}$. Therefore, the sign of $\frac{h_n}{k_n} - \sqrt{d}$ is the same as that of $\frac{h_n}{k_n} - \frac{h_{n+1}}{k_{n+1}}$. By Lemma 5.5, this sign is $(-1)^{n-1}$. On the other hand, $h_n + k_n\sqrt{d}$ is positive, so we must have

$$h_n^2 - d \cdot k_n^2 = (h_n + k_n\sqrt{d})(h_n - k_n\sqrt{d}) = (-1)^{n-1}.$$

For some positive real number α, we have

$$\sqrt{d} = [a_0, a_1, \ldots, a_n, \alpha].$$

We'll solve this equation and find α. Proposition 5.3 implies

$$\sqrt{d} = \frac{\alpha h_n + h_{n-1}}{\alpha k_n + k_{n-1}}.$$

This gives us the equation:

$$(h_n - k_n\sqrt{d})\alpha = -(h_{n-1} - k_{n-1}\sqrt{d}).$$

Multiplying by $h_n + k_n\sqrt{d}$, we get

$$\alpha = (-1)^n(h_{n-1} - k_{n-1}\sqrt{d})(h_n + k_n\sqrt{d}).$$

Expanding out and using the fact that $h_n k_{n-1} - k_n h_{n-1} = (-1)^{n-1}$, we get

$$\alpha = c + \sqrt{d}, \quad \text{where } c = (-1)^{n-1}(h_n h_{n-1} - k_n k_{n-1} d).$$

In particular, this shows that the continued fraction expansion of $\sqrt{d}$ is eventually periodic:

$$\sqrt{d} = [a_0, \ldots, a_n, c + \sqrt{d}] = [a_0, \overline{a_1, \ldots, a_n, c + a_0}].$$

Conversely, assume that $\sqrt{d} = [a_0, \overline{a_1, \ldots, a_n, c + a_0}]$ for some integer c; we'll prove that $\mathrm{N}(h_n + k_n\sqrt{d}) = (-1)^{n-1}$ and $c = a_0$. Our assumption implies $\sqrt{d} = [a_0, a_1, \ldots, a_n, c + \sqrt{d}]$, and hence

$$\sqrt{d} = \frac{(c + \sqrt{d})h_n + h_{n-1}}{(c + \sqrt{d})k_n + k_{n-1}}.$$

We'll express the right-hand side in the form $x + y\sqrt{d}$ and then compare coefficients. Multiplying the numerator and denominator by $(c - \sqrt{d})k_n + k_{n-1}$, we get

$$\sqrt{d} = \frac{(ch_n + \sqrt{d}h_n + h_{n-1})(ck_n - \sqrt{d}k_n + k_{n-1})}{N},$$

where $N = \mathrm{N}(ck_n + k_{n-1} + \sqrt{d}k_n)$. Expanding out the brackets in the numerator and cancelling, we get

$$\sqrt{d} = \frac{(ch_n + h_{n-1})(ck_n + k_{n-1}) - dh_n k_n + \sqrt{d}(h_n k_{n-1} - k_n h_{n-1})}{N}.$$

Comparing coefficients of $\sqrt{d}$, we get $N = h_n k_{n-1} - k_n h_{n-1} = (-1)^{n-1}$. By Proposition 5.7, it follows that the fraction $\frac{ck_n + k_{n-1}}{k_n}$ is a convergent in the continued fraction expansion of $\sqrt{d}$, and therefore $ck_n + k_{n-1} = h_n$. In particular, $N = \mathrm{N}(h_n + k_n\sqrt{d}) = (-1)^{n-1}$.

The equation $ck_n + k_{n-1} = h_n$ implies $c = \frac{h_n}{k_n} - \frac{k_{n-1}}{k_n}$, and in particular

$$\frac{h_n}{k_n} - 1 < c < \frac{h_n}{k_n}.$$

Since $\lfloor\sqrt{d}\rfloor \leq \frac{h_n}{k_n} \leq \lfloor\sqrt{d}\rfloor + 1$, it follows that

$$\lfloor\sqrt{d}\rfloor - 1 < c < \lfloor\sqrt{d}\rfloor + 1.$$

Since c is an integer, we must have $c = \lfloor\sqrt{d}\rfloor = a_0$. Therefore,

$$\sqrt{d} = [a_0, \overline{a_1, \ldots, a_n, 2a_0}].$$

□

Corollary 5.4. *Let $d \geq 2$ be square-free. The continued fraction expansion of $\sqrt{d}$ is eventually periodic with the form $[a_0, \overline{a_1, \ldots, a_n, 2a_0}]$. There is an element in $\mathbb{Z}[\sqrt{d}]$ with norm -1 if and only if n is even.*

In fact, one can show a little more than this. It turns out that the partial quotients $a_1, \ldots, a_n$ remain the same if we reverse their order; $a_{n+1-r} = a_r$ for $r = 1, 2, \ldots, n$. This is proved in [7].

A little earlier we considered the case $d = 7$, and showed that $(8, 3)$ is the fundamental solution to the equation $x^2 - 7y^2 = 1$. We could have found this fundamental solution by hand since the numbers are small. We'll look at another example where the fundamental solution is a little bigger, so the continued fraction method is much more useful.

Example. Consider Pell's equation in the case $d = 13$:

$$x^2 - 13y^2 = 1.$$

After a short calculation, we find that the continued fraction expansion of $\sqrt{13}$ is

$$\sqrt{13} = [3, \overline{1, 1, 1, 1, 6}].$$

The periodic section has odd length, and so there is an element of $\mathbb{Z}[\sqrt{13}]$ of norm -1. To find the smallest such element, we calculate

$$[3, 1, 1, 1, 1] = \frac{18}{5}.$$

We can easily check that $\mathrm{N}(18+5\sqrt{13}) = 18^2 - 13\cdot 5^2 = -1$. To find the fundamental solution to Pell's equation, we calculate:

$$[3,1,1,1,1,6,1,1,1,1] = \frac{649}{180}.$$

Therefore, $\mathrm{N}(649+180\sqrt{13}) = 1$. In fact, $649+180\sqrt{13} = (18+5\sqrt{13})^2$.

Units in real quadratic rings. There are slight differences between solving Pell's equation and finding units in real quadratic rings. Units may have norm either 1 or -1, but the solutions to Pell's equation only give units of norm 1. Furthermore, in the case $d \equiv 1 \bmod 4$, we may have units in $\mathbb{Z}[\alpha]$ which are not in the subring $\mathbb{Z}[\sqrt{d}]$. We'll see next how to find all the units. We saw in Proposition 5.7 that if $x+y\sqrt{d}$ is a unit in $\mathbb{Z}[\sqrt{d}]$ with x and y positive, then $\frac{x}{y}$ is a convergent in the continued fraction expansion of $\sqrt{d}$. The following result is similar, but applies to units in the ring $\mathbb{Z}[\alpha]$, which is sometimes bigger.

Proposition 5.8. *Let $x - y\alpha \in (-1,1)$ be a unit in $\mathbb{Z}[\alpha]$ with $y > 0$. Then, $\frac{x}{y}$ is a convergent in the continued fraction expansion of α.*

Proof. Let $A = x - y\alpha$. Since $|A| < 1$, it follows that $x > y\alpha - 1$ and in particular x is also positive. As A is a unit, we have $|x - y\alpha|\cdot|x - y\bar{\alpha}| = 1$. Therefore,

$$\left|\frac{x}{y} - \alpha\right| = \frac{1}{|y|\cdot|x - y\bar{\alpha}|}.$$

Since $x > y\alpha - 1$, we have $x - y\bar{\alpha} > \delta y - 1$ where $\delta = \alpha - \bar{\alpha}$. The number δ is either $\sqrt{d}$ or $2\sqrt{d}$, depending on whether or not d is congruent to 1 modulo 4.

Suppose for a moment that $y > \frac{1}{\delta - 2}$. This implies $\delta y - 1 > 2y$, and therefore

$$\left|\frac{x}{y} - \alpha\right| < \frac{1}{2y^2}.$$

It follows by Theorem 5.6, that $\frac{x}{y}$ is a convergent to α.

It remains to consider the cases where $1 \le y < \frac{1}{\delta - 2}$. In particular, we have $\delta < 3$, which only happens in the cases $d = 2$ or $d = 5$. In the case $d = 2$, the inequality $1 \le y < \frac{1}{2\sqrt{2}-2}$ implies $y = 1$. Since $A = x - \sqrt{2}$ is a unit, we can easily check that $x = 1$. The result is true in this case because $\frac{1}{1}$ is a convergent of $\sqrt{2}$.

In the case $d = 5$, the inequality $1 \le y < \frac{1}{\sqrt{5}-2}$ means $1 \le y < 5$. There are four units A satisfying this condition:

$$1 - \alpha, \;\; 2 - \alpha, \;\; 3 - 2\alpha, \;\; 5 - 3\alpha.$$

We may check that all of the fractions $\frac{1}{1}$, $\frac{2}{1}$, $\frac{3}{2}$ and $\frac{5}{3}$ are convergents to $\frac{1+\sqrt{5}}{2}$. Indeed, the continued fraction expansion of $\frac{1+\sqrt{5}}{2}$ is $[1, 1, 1 \ldots]$, and the convergents are the ratios of successive Fibonacci numbers (see Exercise 5.21). Therefore, the theorem is true in these cases as well. □

Exercise 5.26.

(1) Find the continued fraction expansion of each of the following:

$$\sqrt{2}, \quad \sqrt{3}, \quad \sqrt{5}, \quad \sqrt{6}, \quad \sqrt{10}.$$

(you can check your answers by typing `continued_fraction(sqrt(2))` into `sage`.)

(2) For each for the following integers d, find the fundamental solution to Pell's equation $x^2 - dy^2 = 1$.

$$2, \; 3, \; 5, \; 6, \; 10.$$

(3) Hence, find two solutions in positive integers to the equation

$$x^2 - 10y^2 = 26.$$

5.7 REAL QUADRATIC RINGS AND DIOPHANTINE EQUATIONS

In this section, we'll solve the following Diophantine equation:

$$x^3 = y^2 - 2. \tag{5.8}$$

The difference between this equation and the equations considered earlier in this chapter, is that the right-hand side $y^2 - 2$ factorizes as $(y + \sqrt{2})(y - \sqrt{2})$ over a real quadratic ring $\mathbb{Z}[\sqrt{2}]$, rather than over a complex quadratic ring. As we saw in Section 5.6, the ring $\mathbb{Z}[\sqrt{2}]$ has infinitely many units, so it is more complicated to apply Lemma 5.1 in this case.

As before, we begin by showing that the factors $y + \sqrt{2}$ and $y - \sqrt{2}$ are coprime in $\mathbb{Z}[\sqrt{2}]$. If D were a common factor, then D would be a factor, of $2\sqrt{2}$, and therefore of 4. However, x must be odd, since otherwise $y^2 - 2 \equiv 0 \bmod 8$. Since D is a common factor of x and 4, it follows that D is a unit.

Theorem 5.2 tells us that $\mathbb{Z}[\sqrt{2}]$ has unique factorization. Therefore, by Lemma 5.1, there is a unit U in $\mathbb{Z}[\sqrt{2}]$ such that

$$y + \sqrt{2} = UV^3, \quad V \in \mathbb{Z}[\sqrt{2}].$$

We've seen in the previous section how to find the units in $\mathbb{Z}[\sqrt{2}]$. They all have the form $U = \pm F^n$, where $F = 1 + \sqrt{2}$. Let's write $n = 3m + r$ where r is 0, 1 or -1. Replacing V by $\pm F^m V$, we have a solution to one of the three equations:

$$y + \sqrt{2} = (1 + \sqrt{2})^r (h + k\sqrt{2})^3, \quad r = 0,\ 1, \text{ or } -1.$$

The case $r = 0$ is rather similar to some examples which we have considered earlier. Expanding out the right-hand side, we have

$$y + \sqrt{2} = (h + k\sqrt{2})^3 = (h^3 + 2hk^2) + \sqrt{2}(3h^2k + 2k^3).$$

Equating coefficients of $\sqrt{2}$, we get $k(3h^2 + 2k^2) = 1$. This equation has no solutions in integers.

The case $r = 1$ is rather different. If we expand out in this case, then we have

$$y + \sqrt{2} = (h^3 + 6h^2k + 6hk^2 + 4k^3) + \sqrt{2}(h^3 + 3h^2k + 6hk^2 + 2k^3).$$

Equating coefficients of $\sqrt{2}$ as before, we have

$$h^3 + 3h^2k + 6hk^2 + 2k^3 = 1. \tag{5.9}$$

Equation (5.9) is a much harder equation to solve than we are used to, because it does not factorize. Before solving this equation, we'll examine the final case $r = -1$. In this case, we have

$$y + \sqrt{2} = F^{-1} \cdot V^3.$$

Since $\mathrm{N}(F) = -1$, the conjugate of F is $-F^{-1}$. Therefore, if we take the conjugate of the equation above, we get

$$-y + \sqrt{2} = F \cdot W^3, \quad \text{where } W = \bar{V}.$$

This equation has a solution with a given integer $y = c$ if and only if the previous equation $y + \sqrt{2} = F \cdot V^3$ has a solution with $y = -c$. It is therefore sufficient to find all solutions to Equation (5.9).

The continued fraction method. We'll now try to solve Equation (5.9). There is an obvious solution $(h, k) = (1, 0)$. Substituting these values, we find $y = 1$, which gives us a solution $(x, y) = (-1, 1)$ to Equation (5.8). In the case $r = -1$, we have the corresponding solution $(x, y) = (-1, -1)$.

Let's suppose that (h, k) is any other solution to Equation (5.9), so that we have $k \neq 0$. Dividing by k^3 we get

$$f\left(\frac{h}{k}\right) = \frac{1}{k^3}, \quad \text{where } f(X) = X^3 + 3X^2 + 6X + 2.$$

It's worth sketching the graph of f.

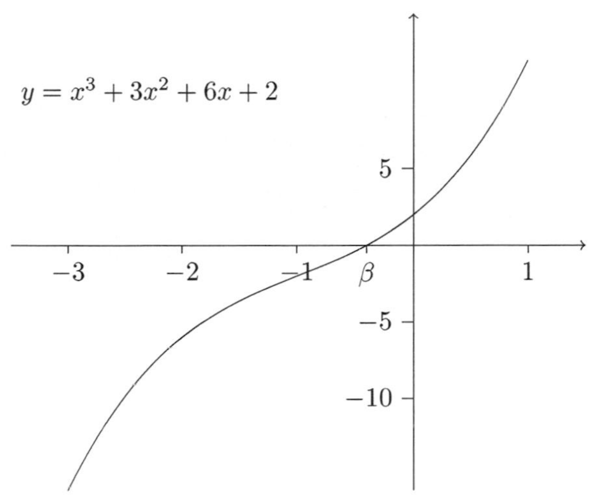

From the graph, we see that f has just one real root β, which is an irrational number between -1 and 0. Furthermore, for all $x \in \mathbb{R}$, we have $f'(x) = 3x^2 + 6x + 6 \geq 3$. By the Mean Value Theorem, this implies

$$|f(x)| \geq 3|x - \beta| \quad \text{for all } x \in \mathbb{R}.$$

We've shown that the integers h and k satisfy $|f(\frac{h}{k})| = \frac{1}{|k|^3}$, so the inequality above implies

$$\left|\frac{h}{k} - \beta\right| \leq \frac{1}{3|k|^3}.$$

Since $\frac{1}{3|k|^3} < \frac{1}{2k^2}$, Theorem 5.6 shows that $\frac{h}{k}$ must be one of the convergents in the continued fraction expansion of β. This allows us to search for solutions to Equation (5.9) in the same way that we searched for solutions to Pell's equation.

We'll consider how hard it is to calculate the continued fraction expansion of β. We calculate a sequence of real numbers β_n and integers a_n, which are defined by the recursive formulae:

$$\beta_0 = \beta, \quad a_n = \lfloor \beta_n \rfloor, \quad \beta_{n+1} = \frac{1}{\beta_n - a_n}.$$

The sequence β_n is contained in the field $\mathbb{Q}(\beta) = \{x + y\beta + z\beta^2 : x, y, z \in \mathbb{Q}\}$. The only difficulty is in calculating with complete accuracy the floor of an element of $\mathbb{Q}(\beta)$. The floor of an element $\gamma \in \mathbb{Q}(\beta)$ is the largest integer n, such that $\mathrm{N}(\gamma - n) > 0$, where the norm $\mathrm{N}(x + y\beta + z\beta^2)$ is defined to be the determinant of the matrix

$$\begin{pmatrix} x & y & z \\ -2z & x - 6z & y - 3z \\ -2y + 6z & -6y + 16z & x - 3y + 3z \end{pmatrix}.$$

This is because one can check that $\mathrm{N}(x + y\beta + z\beta^2)$ is also equal to the product $(x + y\beta + z\beta^2)|x + y\delta + z\delta^2|^2$, where δ is one of the complex roots of f. Using this method, we can calculate several convergents of β, in order to search for a solution to Equation (5.9). The first few partial quotients in the continued fraction expansion of β are

$$\beta = [-1, 1, 1, 2, 9, 1, 3, 1, 1, 2, 7, 4, 5, 2, 2, 2, 4, \ldots].$$

The corresponding convergents $\frac{h}{k}$ together with the values of $k^3 f(\frac{h}{k})$ are given in the following table.

$\frac{h}{k}$	$h^3 + 3h^2k + 6hk^2 + 2k^3$
-1	-2
0	2
$-\frac{1}{2}$	-3
$-\frac{2}{5}$	2
$-\frac{19}{47}$	-138
$-\frac{21}{52}$	47
$-\frac{82}{203}$	-426
$-\frac{103}{255}$	458
$-\frac{185}{458}$	-691
$-\frac{473}{1171}$	624
$-\frac{3496}{8655}$	-8146
$-\frac{14457}{35791}$	25724
$-\frac{75781}{187610}$	-293511
$-\frac{166019}{411011}$	574922
$-\frac{407819}{1009632}$	-1562003
$-\frac{981657}{2430275}$	2000032
$-\frac{4334447}{10730732}$	-20475691

If $(\pm h, \pm k)$ were a solution to Equation (5.9), then the number in the right-hand column would be ± 1. We see that there is no ± 1 in the right-hand column, so we have found no new solution to Equation (5.9). In particular, this shows that in any solution apart from the obvious solution $(1, 0)$, we must have at least $|k| > 10730732$. Although the continued fraction method is a very fast way of searching for solutions, the table might lead one to think that there are no more solutions.

Thue equations. A Thue equation is a Diophantine equation of the form,

$$a_0 x^d + a_1 x^{d-1} y + \cdots + a_d y^d = c, \tag{5.10}$$

where the coefficients a_i and c are given integers, and the degree d is at least 3. It was shown by Thue that there are only finitely many integer solutions (x, y) to Equation (5.10). In particular, this shows that Equation (5.9) has only finitely many solutions, and hence Equation (5.8) has only finitely many solutions. However, Thue's theorem does not give us a way of checking that we have found all the solutions. Baker produced an explicit bound on the size of the solutions to Equation (5.10) (see [3]). Baker's bound was later improved to such an extent that one can now solve a Thue equation such as Equation (5.9) on a computer. In fact, the following code shows that there are no solutions apart from the solution $(h, k) = (1, 0)$, which we noticed earlier.

```
%gp
t=thueinit(x^3+3*x^2+6*x+2)
thue(t,1)
      [[1,0]].
```

Once `t` is initialized as above, the command `thue(t,c)` will find all integer solutions to $x^3+3x^2y+6xy^2+2y^3 = c$. This shows that $(-1, 1)$ and $(-1, -1)$ are the only solutions in integers to Equation (5.8).

Skolem's *p*-adic method. Using a computer to solve the equation seems like cheating, and there is another older method called *Skolem's p-adic method* which we'll now sketch out. Skolem's method is to use congruences in the ring $\mathbb{Z}[\beta]$, where β is the root of f. Since f is monic and irreducible, β is an algebraic integer and the elements of $\mathbb{Z}[\beta]$ all have the form $a_0 + a_1\beta + a_2\beta^2$ with integer coefficients a_0, a_1, a_2. By the remainder theorem, $X - \beta$ is a factor of f in the ring of polynomials over $\mathbb{Z}[\beta]$. Indeed, dividing f by $X - \beta$, we find

$$f(X) = (X - \beta)(X^2 + \gamma X + \delta), \quad \text{where } \gamma = 3 + \beta, \;\; \delta = \beta^2 + 3\beta + 6.$$

Suppose (h,k) is any solution in integers to Equation (5.9). This implies

$$(h-k\beta)(h^2+\gamma hk+\delta k^2)=1,$$

and in particular, $h-k\beta$ is a unit in the ring $\mathbb{Z}[\beta]$.

Rather like the real quadratic rings, the ring $\mathbb{Z}[\beta]$ has infinitely many units. The units in $\mathbb{Z}[\beta]$ all have the form $\pm G^n$ for a particular unit G; in this case, $G=-\beta^2-3\beta-1$. If we can find all the integers n, such that $\pm G^n$ has the form $h-k\beta$, then the list of pairs (h,k) must include all solutions to Equation (5.9). The following lemma shows that such an n must be 0. This implies that $(h,k)=(1,0)$ is the only solution to Equation (5.9).

Lemma 5.7. *Let β and G be as above. If $G^n=h-k\beta$ with $h,k\in\mathbb{Z}$, then $n=0$.*

Sketch proof. Here are the first few powers of G:

n	G^n	$G^n \bmod 3\mathbb{Z}[\beta]$
0	1	1
1	$-\beta^2-3\beta-1$	$2+2\beta^2$
2	$-4\beta^2-14\beta-5$	$1+2\beta+2\beta^2$
3	$-15\beta^2-63\beta-23$	1

In particular, we see from the table that $G^3\equiv 1 \bmod 3\mathbb{Z}[\beta]$. Therefore, the congruency class of G^n modulo $3\mathbb{Z}[\beta]$ depends only on n modulo 3. Furthermore, one can see from the table that unless n is a multiple of 3, the coefficient of β^2 in G^n is non-zero modulo 3. This shows that if $G^n=h-k\beta$, then n must be a multiple of 3.

It remains to show that if $m\neq 0$, then G^{3m} does not have the form $h-k\beta$. We'll calculate the congruency classes of G^{3m} modulo powers of 3 in the ring $\mathbb{Z}[\beta]$.

$$G^3=1-3(8+21\beta+5\beta^2).$$

In particular, the logarithm $L=\log(G^3)$ converges 3-adically, and modulo 9 is given by

$$L\equiv -3(8+21\beta+5\beta^2)\equiv 3+3\beta^2 \mod 9\mathbb{Z}[\beta]. \tag{5.11}$$

Because of the convergence of this logarithm, we can expand G^{3m} modulo any power of 3 as a 3-adic power series in the variable m:

$$G^{3m}\equiv \exp(mL) \mod 3^N\mathbb{Z}[\beta].$$

We shall calculate the power series modulo 3^N, where $N = v_3(m) + 2$. One can easily show that modulo this power of 3, only the first two terms of the power series are non-zero, so we have

$$G^{3m} \equiv 1 + mL \mod 3^N\mathbb{Z}[\beta].$$

In particular, from Equation (5.11), we have

$$G^{3m} \equiv 1 + 3m + 3m\beta^2 \mod 3^N\mathbb{Z}[\beta].$$

Since $3m \not\equiv 0 \bmod 3^N$, it follows that the coefficient of β^2 in G^{3m} is non-zero. □

We have shown that if $h - k\beta$ is a unit in $\mathbb{Z}[\beta]$ then $h - k\beta = \pm 1$, and therefore the only solution to Equation (5.9) is the obvious solution $(1, 0)$.

HINTS FOR SOME EXERCISES

5.8. In each case, we find Q first; Q is the nearest element in the quadratic ring to $\frac{A}{B}$.

5.12. Use the method described in Section 5.1, together with the list of complex quadratic rings with unique factorization.

5.14. The ring $\mathbb{Z}[\frac{1+\sqrt{5}}{2}]$ has unique factorization.

5.16. Prove the first part by induction on $|\mathrm{N}(A)|$. For the second part, let $N = |\mathrm{N}(P_1 \cdots P_r)| + 1$.

5.25. Write Z as $\frac{A}{B}$ with $A, B \in \mathbb{Z}[\alpha]$, and prove by induction on $|\mathrm{N}(B)|$.

5.26. In the last part, find one solution, and then use the fundamental solution to Pell's equation to get another.

Solution to Exercises

1.1. Let $x - x' = na$ and $y - y' = nb$. Then we have $(x + y) - (x' + y') = n(a + b)$ and $xy - x'y' = n(xb + ay')$.

1.2. We use Euclid's algorithm.

$$\begin{aligned}
\mathbf{136} &= 1 \cdot \mathbf{123} + \mathbf{13} & 1 &= \mathbf{13} - 2 \cdot \mathbf{6} \\
\mathbf{123} &= 9 \cdot \mathbf{13} + \mathbf{6} & &= \mathbf{13} - 2(\mathbf{123} - 9 \cdot \mathbf{13}) \qquad = 19 \cdot \mathbf{13} - 2 \cdot \mathbf{123} \\
\mathbf{13} &= 2 \cdot \mathbf{6} + \mathbf{1} & &= 19(\mathbf{136} - \mathbf{123}) - 2 \cdot \mathbf{123} = 19 \cdot \mathbf{136} - 21 \cdot \mathbf{123}.
\end{aligned}$$

Therefore, $h = 19$, $k = -21$ is a solution.

1.3. Suppose $a = rn + sm$. Any common factor of n and m must also be a factor of a. Therefore, $\mathrm{hcf}(n, m)|a$. Conversely, if $a = b \cdot \mathrm{hcf}(n, m)$, then $a = bhn + bkm$, where $hn + km = \mathrm{hcf}(n, m)$.

1.4. If d is a factor of $\mathrm{hcf}(n, m)$, then it's clear that $d|n$ and $d|m$. This follows from the fact that $a|b$ and $b|c$ imply $a|c$. Conversely, assume that $d|n$ and $d|m$. Using the equation $\mathrm{hcf}(n, m) = hn + km$, we see that d must be a factor of $\mathrm{hcf}(n, m)$.

1.5. Let $\mathrm{lcm}(n, m) = na = mb$ and choose h, k such that $\mathrm{hcf}(n, m) = hn + km$. This implies $\mathrm{lcm}(n, m)\mathrm{hcf}(n, m) = nm(hb + ka)$, which is a multiple of nm. Therefore, $\mathrm{lcm}(n, m)\mathrm{hcf}(n, m) \geq nm$. On the other hand, $\frac{nm}{\mathrm{hcf}(n,m)}$ is a common multiple of n and m, because both $\frac{n}{\mathrm{hcf}(n,m)}$ and $\frac{m}{\mathrm{hcf}(n,m)}$ are integers. This implies $\mathrm{lcm}(n, m) \leq \frac{nm}{\mathrm{hcf}(n,m)}$. The two inequalities together show that $\mathrm{lcm}(n, m)\mathrm{hcf}(n, m) = nm$

1.6. The integers 1, 5, 7 and 11 are coprime to 12, whereas the even integers and the multiples of 3 are not. Therefore, $(\mathbb{Z}/12)^\times = \{1, 5, 7, 11\}$.

Multiplication is given by

$\times$	1	5	7	11
1	1	5	7	11
5	5	1	11	7
7	7	11	1	5
11	11	7	5	1

1.7. We use Euclid's algorithm to show that 175 and 9 are coprime.

$$\begin{aligned} 175 &= 19 \times 9 + 4 & \qquad 1 &= 9 - 2 \times 4 \\ 9 &= 2 \times 4 + 1 & &= 9 - 2 \times (175 - 19 \times 9) \\ & & &= 39 \times 9 - 2 \times 175. \end{aligned}$$

Reducing the last equation modulo 175, we have $1 \equiv 39 \times 9 \bmod 175$. Therefore, $9^{-1} \equiv 39 \bmod 175$. The same method shows that $23^{-1} \equiv -38 \equiv 137 \bmod 175$:

$$\begin{aligned} 175 &= 7 \times 23 + 14 \quad 1 = 5 - 4 \\ 23 &= 1 \times 14 + 9 = 5 - (9 - 5) = 2 \times 5 - 9 \\ 14 &= 1 \times 9 + 5 = 2(14 - 9) - 9 = 2 \times 14 - 3 \times 9 \\ 9 &= 1 \times 5 + 4 = 2 \times 14 - 3 \times (23 - 14) = 5 \times 14 - 3 \times 23 \\ 5 &= 1 \times 4 + 1 = 5 \times (175 - 7 \times 23) - 3 \times 23 = 5 \times 175 - 38 \times 23. \end{aligned}$$

1.8. Suppose x is a solution to the congruence. Then there is an integer y such that $ax + ny = b$. Since a is a factor of n, it follows that a is a factor of b, which gives a contradiction.

1.9. The first congruence implies there is an integer y such that $ax+ny = b$. This is equivalent to $x + \frac{n}{a}y = \frac{b}{a}$, which is the second congruence.

1.10. $x \equiv 54 \bmod 71$, $x \equiv 2 \bmod 71$, no solutions, $x \equiv 15 \bmod 18$, $x \equiv 3 \bmod 11$.

1.11. (a) $x \equiv 19 \bmod 40$, (b) $x \equiv 77 \bmod 475$.

1.12. Let $d = \mathrm{hcf}(n, m)$. Suppose that a solution x exists. Then we have

$$a \equiv x \equiv b \mod d.$$

Conversely, suppose that $a \equiv b \bmod d$, so $(b-a)/d$ is an integer. Since n/d and m/d are coprime integers, there is a solution y to the simultaneous congruences

$$y \equiv 0 \bmod n/d, \quad y \equiv (b-a)/d \bmod m/d.$$

Then we can check that $x = a + dy$ is a solution to the original pair of congruences.

1.13. The set $R \times S$ with the operation of addition is the direct sum of the groups $(R, +)$ and $(S, +)$ and so $(R \times S, +)$ is an abelian group. We'll check that multiplication is commutative and associative and has an identity element $(1, 1)$:

$$\big((x_1, y_1)(x_2, y_2)\big)(x_3, y_3) = (x_1x_2x_3, y_1y_2y_3) = (x_1, y_1)\big((x_2, y_2)(x_3, y_3)\big),$$

$$(x_1, y_1)(x_2, y_2) = (x_1x_2, y_1y_2) = (x_2, y_2)(x_1, y_1),$$

$$(1, 1)(x, y) = (1x, 1y) = (x, y).$$

Finally, we check the distributivity axiom:

$$\begin{aligned}(x_1, y_1)\big((x_2, y_2) + (x_3, y_3)\big) &= (x_1(x_2 + x_3), y_1(y_2 + y_3)) \\ &= (x_1x_2 + x_1x_3, y_1y_2 + y_1y_3) \\ &= (x_1, y_1)(x_2, y_2) + (x_1, y_1)(x_3, y_3).\end{aligned}$$

We already know that $\mathcal{C}$ is bijective. To see that the map $\mathcal{C}$ is a ring isomorphism, we need to show that $\mathcal{C}(x + y) = \mathcal{C}(x) + \mathcal{C}(y)$ and $\mathcal{C}(xy) = \mathcal{C}(x)\mathcal{C}(y)$:

$$\begin{aligned}\mathcal{C}(x + y) &= (x + y \bmod n, x + y \bmod m) \\ &= (x \bmod n, x \bmod m) + (y \bmod n, y \bmod m), \\ \mathcal{C}(xy) &= (xy \bmod n, xy \bmod m) \\ &= (x \bmod n, x \bmod m)(y \bmod n, y \bmod m).\end{aligned}$$

1.14. The multiplication table is

$\times$	1	2	3	4	5	6
1	1	2	3	4	5	6
2	2	4	6	1	3	5
3	3	6	2	5	1	4
4	4	1	5	2	6	3
5	5	3	1	6	4	2
6	6	5	4	3	2	1

The inverses are $1^{-1} \equiv 1$, $2^{-1} \equiv 4$, $3^{-1} \equiv 5$, $4^{-1} \equiv 2$, $5^{-1} \equiv 3$ and $6^{-1} \equiv 6 \bmod 7$.

1.15. If $n = 1$, then $1 \equiv 0 \bmod n$ so $\mathbb{Z}/n$ is not a field. If n is composite, then $n = rs$ with $r, s > 1$. This implies r and s are both non-zero modulo n. If $\mathbb{Z}/n$ were a field, then r would have an inverse, and so $rr^{-1} \equiv 1 \bmod n$. This implies $0 \equiv r^{-1} \cdot 0 \equiv r^{-1}n \equiv r^{-1}rs \equiv s$, which gives a contradiction.

1.16. Our conditions on S imply that addition and multiplication restrict to give binary operations on S. The ring axioms are equations which hold for all elements of the whole ring R. In particular, they are also true for all elements of S.

1.17. The primes up to 30 are 2, 3, 5, 7, 11, 13, 17, 19, 23 and 29.

$$323 = 17 \times 19, \quad 329 = 7 \times 47, \quad 540 = 2^2 \times 3^2 \times 5, \quad 851 = 23 \times 37.$$

The other numbers 263 and 617 in the list are prime.

1.18. Suppose there are only finitely many, and call them $p_1, \ldots, p_r$. Let $N = 5p_1 \cdots p_r - 1$. Clearly, $N \equiv 4 \bmod 5$. However, N is not a multiple of any of the primes $p_1, \ldots, p_r$. Therefore, N is a product of primes which are congruent to 1 modulo 5, and this implies $N \equiv 1 \bmod 5$. This gives a contradiction.

1.19. Suppose there are only finitely many primes whose congruency class is in $(\mathbb{Z}/n)^\times \setminus H$. Call these primes $p_1, \ldots, p_r$. Choose a positive integer a whose congruency class modulo n is in $(\mathbb{Z}/n)^\times \setminus H$, and such that $a \equiv 1 \bmod p_1 \cdots p_r$ (such an a exists by the Chinese Remainder Theorem). Let $N = np_1 \cdots p_r + a$.

By our choice of a, we know that $N \equiv 1 \bmod p_i$, so none of the primes p_i are factors of N. Also N is coprime to n. It follows that all the prime factors of N are congruent modulo n to elements of H. Since H is a subgroup, it follows that N is congruent modulo n to an element of H. This implies a is congruent modulo n to an element of H, which gives a contradiction.

1.20. If $p \equiv 210 \bmod 1477$, then p is a multiple of 7. Since p is prime, it follows that $p = 7$, but this gives a contradiction.

1.21. Suppose that there are only finitely many such primes and call them $p_1, \ldots, p_r$. Let $N = (p_1 \cdots p_r)^5 + (p_1 \cdots p_r) + 1$ and let q be a prime factor of N. None of the primes $p_1, \ldots, p_r$ are factors of N. Therefore, q is not one of the primes $p_1, \ldots, p_r$, and so the congruence $x^5 + x + 1 \equiv 0 \bmod q$ has no solutions. However, if we set $x = p_1 \cdots p_r$, then we have $x^5 + x + 1 = N \equiv 0 \bmod q$. This gives a contradiction.

1.22. Assume $a^b - 1$ is prime. For each factorization $b = rs$, we have a factorization

$$a^b - 1 = (a^r - 1)(1 + a^r + a^{2r} + \cdots + a^{(s-1)r}).$$

If $r < b$, then this implies $a^r - 1$ is a proper factor of the prime number $a^b - 1$, and therefore $a^r - 1 = 1$. This implies $a^r = 2$, so $a = 2$ and $r = 1$. We've shown that the only proper factor of b is 1, so b is prime.

1.23. If k is not a power of 2, then it has an odd factor, so $k = rs$ with r odd. This gives a contradiction because we have a factorization:

$$2^k + 1 = (2^s + 1)(1 - 2^s + 2^{2s} - \cdots + 2^{(r-1)s}).$$

1.24. By uniqueness of factorization, the numbers 2^{400} and 3^{400} are coprime. Therefore, by the Chinese Remainder Theorem, there is a unique solution modulo 6^{400}. Clearly, $x = 5$ is a solution so the general solution is $x \equiv 5 \bmod 6^{400}$.

2.1. $q(X) = X^3 + X^2 + 1$, $r(X) = 1$.

2.2. $f - g$ is a polynomial with infinitely many roots. By Corollary 2.3, $f - g = 0$.

2.3. (1) Let

$$A = \{x \in \mathbb{Z}/nm : f(x) \equiv 0 \bmod nm\},$$

$$B = \{x \in \mathbb{Z}/n : f(x) \equiv 0 \bmod n\},$$

$$C = \{x \in \mathbb{Z}/m : f(x) \equiv 0 \bmod m\}.$$

It's enough to show that there is a bijection from $\mathcal{C} : A \to B \times C$ defined by $\mathcal{C}(x) = (x \bmod n, x \bmod m)$.

Note first that if $x \in A$, then the congruence $f(x) \equiv 0 \bmod nm$ implies $f(x) \equiv 0 \bmod n$ and $f(x) \equiv 0 \bmod m$. Hence, $\mathcal{C}(x)$ is an element of $B \times C$. It remains to be shown that each element of $B \times C$ has a unique preimage in A. If $(b, c) \in B \times C$, then by the Chinese Remainder Theorem, there is a unique $x \in \mathbb{Z}/nm$ such that $x \equiv a \bmod n$ and $x \equiv b \bmod m$. This implies $f(x) \equiv f(a) \equiv 0 \bmod n$ and $f(x) \equiv f(a) \equiv 0 \bmod m$. Hence, by the Chinese Remainder Theorem, $f(x) \equiv 0 \bmod nm$, so $x \in A$. This shows that each element of $B \times C$ has a unique preimage in A. Therefore, $\mathcal{C}$ is a bijection.

(2) Note that there is one solution in $\mathbb{F}_2$, and for any prime $p > 2$ we have $X^2 - 1 = (X+1)(X-1)$, so this polynomial has two roots at 1 and -1 in $\mathbb{F}_p$. By Corollary 2.3, there are no more roots so $a(p) = 2$. Hence, the number of solutions is $a(2) \cdot a(3) \cdot a(5) \cdot a(7) \cdot a(11) \cdot a(13) \cdot a(17) \cdot a(19) = 1 \cdot 2 \cdot 2 \cdot 2 \cdot 2 \cdot 2 \cdot 2 \cdot 2 = 128$.

2.4. To check whether f is a factor of g, we divide g by f with remainder, and check whether the remainder is zero. The long division algorithm takes the same course whether we regard f and g as elements of $\mathbb{F}[X]$ or as elements of $\mathbb{E}[X]$.

2.5. (1) We go through Euclid's algorithm starting with f and g.

$$f(X) = (X^2 + X + 1)g(X) + (X + 1)$$
$$g(X) = (2X + 1)(X + 1) + 2.$$

The last remainder is 2, which is a unit, so f and g are coprime.

(2) Working backwards through Euclid's algorithm, we find

$$\begin{aligned} 2 &= g(X) - (2X+1)(X+1) \\ &= g(X) - (2X+1)(f(X) - (X^2+X+1)g(X)) \\ &= (2X^3 + 3X^2 + 3X + 2)g(X) - (2X+1)f(X). \end{aligned}$$

Dividing both sides of this equation by 2, we find the solution $h(X) = -X - \frac{1}{2}$ and $k(X) = X^3 + \frac{3}{2}X^2 + \frac{3}{2}X + 1$.

(3) Substituting α for X in the expression $1 = h(X)f(X) + k(X)g(X)$, we find (since $f(\alpha) = 0$) $k(\alpha)g(\alpha) = 1$. Therefore, $\frac{1}{g(\alpha)} = k(\alpha) = \alpha^3 + \frac{3}{2}\alpha^2 + \frac{3}{2}\alpha + 1$.

2.6. This is answered by Euclid's algorithm in the ring $\mathbb{F}_5[X]$. The answer is $\mathrm{hcf}(f, g) = 1$, $h(X) = 2X^2 + 4X + 3$, $k(X) = 3X^3 + 2X^2 + 3$.

2.7. We can find polynomials h and k such that $\mathrm{hcf}(f, g) = hf + kg$. From this, we see that if $d|f$ and $d|g$ then $d|\mathrm{hcf}(f, g)$.

2.8. Both versions of the highest common factor are calculated by Euclid's algorithm, which does not depend on the field. More formally, we can prove by induction of $\deg(g)$. If $g = 0$, then the highest common factor is in both cases $c^{-1}f$, where c is the leading coefficient of f. Assume $g \neq 0$ and assume the result for all pairs of polynomials, one of whose degrees is smaller than $\deg g$. Dividing f by g in $\mathbb{F}[X]$, we get $f = qg + r$ with $r \in \mathbb{F}[X]$ of degree less than that of g. By the inductive hypothesis, $\mathrm{hcf}_{\mathbb{E}[X]}(g,r) = \mathrm{hcf}_{\mathbb{F}[X]}(g,r)$. This implies

$$\mathrm{hcf}_{\mathbb{E}[X]}(f,g) = \mathrm{hcf}_{\mathbb{E}[X]}(g,r) = \mathrm{hcf}_{\mathbb{F}[X]}(g,r) = \mathrm{hcf}_{\mathbb{F}[X]}(f,g).$$

2.9. Suppose there are only finitely many monic irreducible polynomials, and call them $p_1,\dots,p_r$. Then $g = p_1 \cdots p_r + 1$ is a monic irreducible polynomial, which is not a multiple of any irreducible polynomial. This gives a contradiction.

2.10. Suppose there are only finitely many such polynomials and call them $p_1,\dots,p_r$, starting with $p_1(X) = X+1$. Let $g(X) = X^2 p_2(X)\cdots p_r(X)+p_1$. We see that g is not a multiple of any of the polynomials $p_1,\dots,p_r$. Also, $g(0) = 1$ and $g'(0) = 1$.

Let's suppose that g factorizes into irreducibles as $g = q_1 \cdots q_s$. Since $g(0) = 1$, it follows that $q_i(0) \neq 0$, and so we have $q_i(0) = 1$. This implies (using the rule for differentiating a product) $g'(0) = \sum_i q_i'(0)$. Therefore, there is at least one irreducible factor q_i such that $q_i'(0) \neq 0$, and therefore $q_i'(0) = 1$.

2.11. Suppose f is an irreducible real polynomial of degree at least 3. By the Remainder Theorem, we know that f has no real roots. However the Fundamental Theorem of Algebra shows that f has a complex root $\alpha = a + ib$ with $b \neq 0$. We'll show that the polynomial $g(X) = X^2 - 2aX + (a^2+b^2)$ is a factor of f. Indeed, if we divide f by g, then the remainder will have degree less than 2, so we have

$$f(X) = q(r)g(X) + (u + vX), \quad u, v \in \mathbb{R}.$$

Note that α is a root of both f and g, so if we substitute $X = \alpha$ in this equation we get

$$u + v\alpha = 0.$$

Since $\alpha \notin \mathbb{R}$, we must have $u = v = 0$. Hence, g is a factor of f, so we have a contradiction.

2.12. The non-zero polynomials of degree ≤ 3 over $\mathbb{F}_2$ are

$$1, \quad X, \quad X+1, \quad X^2, \quad X^2+1, \quad X^2+X, \quad X^2+X+1, \quad X^3,$$

$$X^3+1, \quad X^3+X, \quad X^3+X+1, \quad X^3+X^2, \quad X^3+X^2+1,$$

$$X^3+X^2+X, \quad X^3+X^2+X+1.$$

The irreducible polynomials are X, $X+1$, X^2+X+1, X^3+X+1 and X^3+X^2+1. The others factorize as follows:

$$\begin{aligned}
X^2+1 &\equiv (X+1)^2 && \mod 2,\\
X^2+X &\equiv X(X+1) && \mod 2,\\
X^3+1 &\equiv (X+1)(X^2+X+1) && \mod 2,\\
X^3+X &\equiv X(X+1)^2 && \mod 2,\\
X^3+X^2 &\equiv X^2(X+1) && \mod 2,\\
X^3+X^2+X+1 &\equiv (X+1)^3 && \mod 2.
\end{aligned}$$

2.13. The first polynomial satisfies Eisenstein's Criterion with the prime 3. The second polynomial is irreducible over $\mathbb{F}_2$ since it is a cubic with no roots in $\mathbb{F}_2$. For the third polynomial $f(X) = X^5+X^3+X+1$, first note that the only possible roots in $\mathbb{Z}$ are 1 and -1, since any root must be a factor of the constant term. Since neither 1 nor -1 is a root, the only possible factorization would be of the form $f(X) = (X^3+aX^2+bX+c)(X^2+dX+e)$ with integers a, b, c, d, e. Expanding this out and equating coefficients, we find

$$a+d=0, \quad b+ad+e=1, \quad ae+bd+c=0, \quad ce=1.$$

The first and fourth of these equations imply $d=-a$ and $c=e=\pm1$. If $c=e=1$, then the second equation implies $b=a^2$. Substituting this into the third equation we get $-a^3+a+1=0$, which has no integer solution a. If on the other hand $c=e=-1$, then the second equation gives $b=a^2+2$, and then the third equation gives us $a^3+3a+1=0$, which has no integer solution. In either case, we have a contradiction.

2.14. Over $\mathbb{Q}$ we have $f(X)=(X^2+1)(X^2-2)$. Over $\mathbb{R}$ we have $f(X)=(X^2+1)(X+\sqrt{2})(X-\sqrt{2})$, and over $\mathbb{C}$ we have $f(X)=(X+i)(X-i)(X+\sqrt{2})(X-\sqrt{2})$.

2.15. Assume that f and g have a non-constant common factor d. By the Fundamental Theorem of Algebra, d has a complex root α. Since f and g are multiples of d, it follows that α is a root of both f and g.

Assume that f and g are coprime. We can find polynomials h, k such that $1 = hf + kg$. If f and g have a common root α, then $1 = h(\alpha)f(\alpha)+k(\alpha)g(\alpha)=0$, which gives a contradiction. Hence, f and g have no common root.

2.16. (1) (*Reflexivity*) Since $ab = ba$ in R, it follows that $(a, b) \sim (a, b)$.

(*Symmetry*) Assume $(a, b) \sim (a', b')$. This implies $ab' = ba'$, which is equivalent to $a'b = b'a$. Therefore, $(a', b') \sim (a, b)$.

(*Transitivity*) Assume $(a, b) \sim (a', b')$ and $(a', b') \sim (a'', b'')$. Then we have $ab' = ba'$ and $a'b'' = b'a''$. Multiplying the first equation by b'', we get

$$ab'b'' = ba'b'' = bb'a''.$$

Hence, $b'(ab'' - ba'') = 0$. Since R is an integral domain and $b' \neq 0$, it follows that $ab'' - ba'' = 0$, so $(a, b) \sim (a'', b'')$.

(2) Assume that $\frac{a}{b} = \frac{a'}{b'}$ so we have $ab' = ba'$. We must prove that $(ad + bc, bd) \sim (a'd + b'c, b'd)$ and $(ac, bd) \sim (a'c, b'd)$. These relations are proved as follows:

$$(ad + bc)b'd = adb'd + bcb'd = a'bd^2 + bcb'd = (a'd + b'c)bd,$$

$$acb'd = ba'cd.$$

(3) To show that $\mathbb{F}$ is a field, we first show that $(\mathbb{F}, +)$ is an abelian group. It's trivial to show that $+$ is commutative, $\frac{0}{1}$ is the identity element and $\frac{-a}{b}$ as an additive inverse of $\frac{a}{b}$. For associativity, we simply calculate

$$\left(\frac{a}{b} + \frac{c}{d}\right) + \frac{e}{f} = \frac{ad + bc}{bd} + \frac{e}{f} = \frac{adf + bcf + bde}{bdf} = \frac{a}{b} + \left(\frac{c}{d} + \frac{e}{f}\right).$$

It's clear from the definition that multiplication in $\mathbb{F}$ is commutative and associative and $\frac{1}{1}$ is the identity element. For distributivity, we calculate:

$$\begin{aligned}\frac{a}{b}\left(\frac{c}{d} + \frac{e}{f}\right) &= \frac{a}{b} \cdot \frac{cf + de}{df} = \frac{(ac)f + d(ae)}{bdf} \\ &= \frac{(ac)(bf) + (bd)(ae)}{(bd)(bf)} = \frac{ac}{bd} + \frac{ae}{bf} \\ &= \frac{a}{b}\frac{c}{d} + \frac{a}{b}\frac{e}{f}.\end{aligned}$$

At the start of the second line of calculation, we have multiplied the numerator and denominator by d, which does not change the equivalence class.

(4) We just need to show that the map $a \mapsto \frac{a}{1}$ is a ring homomorphism from R to $\mathbb{F}$. Clearly, this map takes 1 to $\frac{1}{1}$, which is the identity element in $\mathbb{F}$. Furthermore, it follows from the definitions of $+$ and $\times$ in $\mathbb{F}$ that $\frac{a}{1} + \frac{b}{1} = \frac{a+b}{1}$ and $\frac{a}{1} \cdot \frac{b}{1} = \frac{a \cdot b}{1}$. Therefore, the map is a homomorphism. Note that if $\frac{a}{1} = \frac{b}{1}$ then we have $(a, 1) \sim (b, 1)$, which implies $a = b$. Therefore, the map $a \mapsto \frac{a}{1}$ is injective, so R is isomorphic to its image in $\mathbb{F}$.

(5) The assumption that R is an integral domain was used in proving that the relation $\sim$ is transitive.

2.17. Let f and g be non-zero polynomials over R with leading terms aX^r and bX^s respectively. Since R is an integral domain, it follows that $ab \neq 0$. Furthermore, the term abX^{r+s} in the expansion of fg has larger degree than all other terms in fg, so it does not cancel. Therefore, $fg \neq 0$.

2.18. Since R is an integral domain, Exercise 2.16 tells us that R is contained in a field $\mathbb{F}$ whose elements are fractions a/b with $a, b \in R$. As R is a UFD, we may assume that a and b have no irreducible common factors in R.

We first show that if $f = gh$ with $f \in R[X]$ and $g, h \in \mathbb{F}[X]$ then there is a $c \in \mathbb{F}^\times$ such that cg and $c^{-1}h$ are in $R[X]$. This is proved by modifying the proof of the Gauss Lemma. In particular, this result shows that there are two kinds of irreducible element in $R[X]$. They are either irreducible elements of R, regarded as constant polynomials, or they are irreducible in $\mathbb{F}[X]$ and primitive as elements of $R[X]$. Here, we call a polynomial in $R[X]$ primitive if there is no irreducible element of R which divides all its coefficients.

Suppose p is an irreducible element of R, and that $p|fg$ in $R[X]$. We must show that either f or g is a multiple of p. If this were not the case, then we could let a_r and b_s be the highest coefficients of f and g respectively which are not multiples of p. From this, we see that the coefficient of X^{r+s} in fg is not a multiple of p, which gives a contradiction (here we use the fact that R is a UFD).

Finally, let p be a primitive polynomial in $R[X]$ which is irreducible over $\mathbb{F}$ and suppose $p|fg$ with $f, g \in R[X]$. Since $\mathbb{F}[X]$ is a UFD, we know that $p|f$ or $p|g$ in $\mathbb{F}[X]$. Without loss of generality, assume $f = ph$ for some $h \in \mathbb{F}[X]$, and we must show that h has coefficients in R. By the Gauss Lemma, there is an $\frac{a}{b} \in \mathbb{F}^\times$ such that both $\frac{a}{b}p$ and $\frac{b}{a}h$ are in $R[X]$. This tells us that b is a factor of each of the coefficients of ap. Since p is primitive, there is no irreducible element of R dividing all the coefficient of p. Therefore (since R is a UFD), every irreducible factor of b in R must be a factor of a. We may therefore assume $b = 1$, so h is in $R[X]$.

3.1. Without loss of generality, $a \leq b$. If $a \equiv b \bmod ord(g)$, then $b = a + \mathrm{ord}(g)c$. Since $g^{\mathrm{ord}(g)} = 1$, we have $g^b = g^a(g^{\mathrm{ord}(g)})^c = g^a$. Conversely, assume $g^a = g^b$ so $g^{b-a} = 1$. We can divide $b - a$ by $\mathrm{ord}(g)$ with remainder:

$$b - a = c \cdot \mathrm{ord}(g) + r, \quad 0 \leq r < \mathrm{ord}(g).$$

It follows that $g^r = g^{b-a} = 1$. However, $\text{ord}(g)$ is the smallest positive integer n such that $g^n = 1$. Therefore, $r = 0$.

3.2. 1 has order 1. 6 has order 2. 2 and 4 have order 3. 3 and 5 have order 6. The orders 1, 2, 3 and 6 are all factors of $|\mathbb{F}_7^\times| = 6$. The group is cyclic because it has an element of order 6.

3.3. The elements are

$$\begin{pmatrix}1 & 0\\0 & 1\end{pmatrix}, \begin{pmatrix}1 & 0\\1 & 1\end{pmatrix}, \begin{pmatrix}1 & 1\\0 & 1\end{pmatrix}, \begin{pmatrix}1 & 1\\1 & 0\end{pmatrix}, \begin{pmatrix}0 & 1\\1 & 0\end{pmatrix}, \begin{pmatrix}0 & 1\\1 & 1\end{pmatrix}.$$

They have orders 1, 2, 2, 3, 2 and 3 respectively. The group is not cyclic (it has no element of order 6; in fact it is not even abelian).

3.4. $3^{6478} \equiv 3^4 \equiv 2 \bmod 79$ (by Fermat's little theorem). $3^{6478} \equiv 1 \bmod 83$ (by Fermat's little theorem). $3^{6478} \equiv 1661 \bmod 6557$ (by the Chinese remainder theorem).

3.5. $x \equiv 4 \bmod 7$, $x \equiv 12 \bmod 17$, $x \equiv 40 \bmod 111$. For the last one of these, we calculate x modulo 3 and modulo 37 and then use the Chinese remainder theorem.

3.6. First note that since p and q are odd, the power is a whole number. We'll show that $x^{(pq-p-q+3)/2} \equiv x \bmod p$. A similar argument proves the congruence modulo q. Then, by the Chinese remainder theorem, the congruence is true modulo pq.

If x is a multiple of p, then both sides of the congruence are 0 modulo p, so let's assume x is not a multiple of p. By Fermat's little theorem, we know $x^{p-1} \equiv 1 \bmod p$. This implies (using the fact that $\frac{q-1}{2}$ is an integer)

$$x^{\frac{pq-p-q+3}{2}} = x^{(p-1)\frac{q-1}{2}+1} \equiv 1^{\frac{q-1}{2}} \cdot x^1 \equiv x \mod p.$$

3.7. The set of all functions $\mathbb{F}_p \to \mathbb{F}_p$ is a vector space over $\mathbb{F}_p$ of dimension p with basis $\{a_0, a_1, \ldots, a_{p-1}\}$, where a_n is defined by

$$a_n(x) = \begin{cases} 1 & x = n, \\ 0 & x \neq n. \end{cases}$$

The vector space of polynomials of degree $\leq p-1$ is also p-dimensional, so it is sufficient to show that each basis vector a_n can be represented as a polynomial. By Fermat's little theorem, we have $a_n(x) \equiv 1 - (x - n)^{p-1} \bmod p$.

3.8. (1) Let $N = \frac{p!}{a!(p-a)!}$. We clearly have $a!(p-a)!N = p!$. Reducing this modulo p, we get $a!(p-a)!N \equiv 0 \bmod p$. None of the numbers $1, \ldots, p-1$ are multiples of p, so by Euclid's lemma $N \equiv 0 \bmod p$.

(2) This follows from the previous part of the question using the binomial theorem.

(3) Clearly, $1^p \equiv 1 \bmod p$. Assume $n^p \equiv n \bmod p$. Then we have

$$(n+1)^p \equiv n^p + 1^p \equiv n+1 \mod p.$$

3.9.

n	1	2	3	4	5	6	7	8	9	10
$\varphi(n)$	1	1	2	2	4	2	6	4	6	4
n	11	12	13	14	15	16	17	18	19	20
$\varphi(n)$	10	4	12	6	8	8	16	6	18	8

3.10. $10000 = 10^4 = 2^4 5^4$. Therefore, $\varphi(10000) = (2-1)2^3(5-1)5^3 = 4000$. By Euler's theorem, $7^{4000} \equiv 1 \bmod 10000$. Therefore,

$$7^{135246872002} \equiv 7^2 \equiv 49 \bmod 10000.$$

For the other part, we use the Chinese remainder theorem (because 65 is not coprime to 10000). Clearly,

$$65^{123456789012345} \equiv 1^{123456789012345} \equiv 1 \bmod 2^4$$

and $65^{123456789012345}$ is certainly a multiple of 5^4 so

$$65^{123456789012345} \equiv 0 \bmod 5^4.$$

Putting these two solutions together by the Chinese remainder theorem, we get

$$65^{123456789012345} \equiv 625 \bmod 10000.$$

3.11. The solutions are $x \equiv 1109 \bmod 2016$, $x \equiv 243 \bmod 2345$, and $x \equiv 128 \bmod 2467$.

The method is the same in each case. Here is how to do the first one. We have $2016 = 2^5 3^2 7$. Therefore, $\varphi(2016) = 16 \times 6 \times 6 = 576$. The solution is $x \equiv 3^{(461^{-1} \bmod 576)} \bmod 2016$. To calculate $461^{-1} \bmod 576$, we use Euclid's algorithm. This gives $461^{-1} \equiv 5 \bmod 576$. Therefore, $x \equiv 5^5 \equiv 1109 \bmod 2016$.

3.12. Let p^a be a power of a prime, such that $\varphi(p^a) \leq B$. Then $p^a \leq 2B$. It follows that there are only finitely many prime powers p^a such that $\varphi(p^a) \leq B$. Let N be the product of all such prime powers. It follows that if $\phi(n) \leq B$ then n is a factor of N.

Alternatively, let $n = 2^s r$ where r is odd, so $\varphi(n) = 2^{s-1}\varphi(r)$. If $\varphi(n) = b$, then $b = 2^{s-1}\varphi(r)$. In particular, $s \leq b$. Also by Euler's theorem, $2^{\varphi(r)} \equiv 1 \bmod r$, and hence $2^b \equiv 1 \bmod r$. This shows that $r \leq 2^b - 1$. We've shown that if $\varphi(n) = b$ then n is one of the finitely many numbers $2^s r$ with $s \leq b$ and $r \leq 2^b - 1$.

3.13. (1) This follows from the formula for a geometric progression.

(2) Using the inequality in the first part, we get

$$\prod_{\text{primes } p<N} \left(\frac{p}{p-1}\right) > \prod_{\text{primes } p<N} \left(1 + \frac{1}{p} + \frac{1}{p^2} + \cdots + \frac{1}{p^N}\right).$$

When we expand out the right-hand side, we get all terms of the form $\frac{1}{n}$, where n is a product of the form $p_1^{a_1} \cdots p_r^{a_r}$ with primes $p_i < N$ and powers $a_i < N$. Every integer $n < N$ may be expressed as such a product, so we have (as well as other terms) the numbers $\frac{1}{n}$ with $n < N$. As all terms in the expansion are positive, we get

$$\prod_{\text{primes } p<N} \left(\frac{p}{p-1}\right) > 1 + \frac{1}{2} + \frac{1}{3} + \cdots + \frac{1}{N}.$$

(3) The series $\sum \frac{1}{n}$ diverges, so we can choose N for which $1 + \frac{1}{2} + \cdots + \frac{1}{N} > \frac{1}{\epsilon}$. By the second part, this implies $\prod_{p<N}(\frac{p-1}{p}) < \epsilon$. If we let n be the product of all the primes $p < N$, then we have $\frac{\varphi(n)}{n} = \prod_{p<N} \frac{p-1}{p} < \epsilon$.

3.14.

prime	primitive roots
5	2, 3
7	3, 5
11	2, 6, 7, 8

3.15. 12, 8, 12. (In general, for a prime p the number of primitive roots is $\varphi(p-1)$.)

3.16. We have $10^d = 1 + nx$ for some integer x. This implies $10^d \cdot \frac{1}{n} = x + \frac{1}{n}$. This means that if we shift the decimal expansion of $\frac{1}{n}$ to the left by d places, then the fractional part will still be exactly $\frac{1}{n}$.

3.17. $(\mathbb{Z}/12)^\times = \{1, 5, 7, 11\}$. Each element apart from 1 has order 2 so there is no generator.

3.18. Note that $\varphi(pq)/2 = (p-1)\frac{q-1}{2}$, which is a multiple of $p-1$. Hence, $x^{\varphi(pq)/2} \equiv 1 \bmod p$ by Fermat's little theorem. Similarly, $x^{\varphi(pq)/2} \equiv 1 \bmod q$. Since p and q are coprime, the Chinese Remainder Theorem tells us that $x^{\varphi(pq)/2} \equiv 1 \bmod pq$. This shows that no element has order bigger than $\varphi(pq)/2$, so there is no generator.

3.19. Recall that g is a primitive root modulo p iff for all primes q dividing $p-1$, $g^{(p-1)/q} \not\equiv 1 \bmod p$.

Let q be a prime dividing $p-1$. Then $(p-1)/q$ is even, so $(-g)^{(p-1)/q} = g^{(p-1)/q} \not\equiv 1 \bmod p$.

3.20. Note that $a^{-n} = 1$ if and only if $a^n = 1$. Hence, a and a^{-1} have the same order. It follows that a is a primitive root if and only if a^{-1} is a primitive root. Also note that a and a^{-1} are only equal if $a = \pm 1$, and ± 1 are not primitive roots unless $p = 2$ or $p = 3$. Hence, for primes bigger than 3, the product of all primitive roots is 1, since for every term a, there is another term a^{-1} which cancels it out. If $p = 2$, then 1 is the only primitive root, so the product is 1, and if $p = 3$, then -1 is the only primitive root, so the product is -1.

3.21. This follows from Proposition 3.3 by taking degrees of both sides of the equation.

3.22. Using Proposition 3.3, we have

$$\Phi_6(X) = \frac{X^6 - 1}{\Phi_1(X)\Phi_2(X)\Phi_3(X)} = \frac{X^6 - 1}{(X^3 - 1)(X + 1)} = X^2 - X + 1,$$

$$\Phi_8(X) = \frac{X^8 - 1}{\Phi_1\Phi_2\Phi_4} = \frac{X^8 - 1}{X^4 - 1} = X^4 + 1,$$

$$\Phi_{10}(X) = \frac{X^{10} - 1}{\Phi_1\Phi_2\Phi_5} = \frac{X^{10} - 1}{(X^5 - 1)(X + 1)}$$

$$= \frac{X^5 + 1}{X + 1} = 1 - X + X^2 - X^3 + X^4.$$

3.23. Since n is odd, we have $\varphi(2n) = \varphi(n)$, so there are the same number of primitive n-th roots of units as primitive $2n$-th roots of unity. It's therefore sufficient to show that if $-\zeta$ has order $2n$ the ζ has order n.

If $-\zeta$ has order $2n$, then $(-\zeta)^n$ is a square root of 1, but is not 1. Therefore, $(-\zeta)^n = -1$ so $\zeta^n = 1$. If ζ has order $d < n$, then $(-\zeta)^{2d} = 1$, which gives a contradiction. Therefore, ζ has order n.

If $n \geq 3$, then $\varphi(n)$ is even, so Φ_n has even degree. This implies

$$\begin{aligned}\Phi_n(-X) &= \prod_{\substack{\text{primitive } n\text{-th} \\ \text{roots of unity } \zeta}} (-X - \zeta) = \prod_{\substack{\text{primitive } 2n\text{-th} \\ \text{roots of unity } \zeta}} (-X + \zeta) \\ &= \prod_{\substack{\text{primitive } 2n\text{-th} \\ \text{roots of unity } \zeta}} (X - \zeta) = \Phi_{2n}(X).\end{aligned}$$

3.24. (1) Suppose ζ is a primitive pn-th root of unity. Then we have $(\zeta^p)^n = 1$, and it's clear that no smaller power of ζ^p is 1, so ζ^p is a primitive n-th root of unity. Conversely, assume that ζ^p has order n. This implies $\zeta^{pn} = 1$ so the order of ζ is a factor of pn. Furthermore, since $\zeta^p \in \langle\zeta\rangle$, it follows that the order of ζ is a multiple of n. Therefore, ζ has order either n or np. If it's order is n, then $(\zeta^p)^{n/p} = 1$, which gives a contradiction. The formula for Φ_{pn} follows straight from the definition of the cyclotomic polynomial, together with the description of the pn-th roots of unity.

(2) Suppose ζ^p has order n. This implies $\zeta^{pn} - 1$ so the order of ζ is a factor of pn. On the other hand, $\zeta^p \in \langle\zeta\rangle$ so the order of ζ is a multiple of n. This shows that ζ has order either n or np.

If ζ has order np, then it's clear that ζ^p must have order n. Suppose ζ has order n. Let a be the inverse of p modulo n. We have $(\zeta^p)^a = \zeta$, and therefore $\langle\zeta\rangle = \langle\zeta^p\rangle$. This implies that ζ^p has the same order as ζ. The formula for Φ_{pn} follows straight from the definition of the cyclotomic polynomial, together with the description of the pn-th roots of unity.

(3) Using the formula in the first part, we have

$$\Phi_{p^a}(1) = \Phi_{p^{a-1}}(1^p) = \cdots = \Phi_p(1).$$

Since $\Phi_p(X) = 1 + X + \cdots + X^{p-1}$, we have $\Phi_{p^a}(1) = p$.

If $n = p^a m$, where $m > 1$ is coprime to p, then using the formulae of the first two parts, we get

$$\Phi_n(1) = \Phi_{pm}(1) = \frac{\Phi_m(1^p)}{\Phi_m(1)}.$$

Since $m > 1$, we know that 1 is not a root of Φ_m, so we have $\Phi_n(1) = 1$.

3.25. The squares modulo 13 are $(\pm 1)^1 = 1$, $(\pm 2)^2 = 4$, $(\pm 3)^2 = 9$, $(\pm 4)^2 \equiv 3$, $(\pm 5)^2 \equiv 12$ and $(\pm 6)^2 \equiv 10$. Hence, the quadratic residues are 1, 3, 4, 9, 10 and 12 with roots 1, 4, 2, 3, 6 and 5, respectively.

3.26. We calculate each of these using the Quadratic Reciprocity Law and the First and Second Nebensatz.

$$\left(\frac{-2}{13}\right) = \left(\frac{-1}{13}\right)\left(\frac{2}{13}\right) = (+1)(-1) = -1,$$

$$\begin{aligned}\left(\frac{96}{149}\right) &= \left(\frac{32}{149}\right)\left(\frac{3}{149}\right) = \left(\frac{2}{149}\right)^5\left(\frac{149}{3}\right)\\ &= (-1)^5\left(\frac{2}{3}\right) = (-1)(-1) = 1,\end{aligned}$$

$$\begin{aligned}\left(\frac{-102}{199}\right) &= \left(\frac{-1}{199}\right)\left(\frac{2}{199}\right)\left(\frac{3}{199}\right)\left(\frac{17}{199}\right)\\ &= (-1)(+1)(-1)\left(\frac{199}{3}\right)\left(\frac{199}{17}\right)\\ &= \left(\frac{1}{3}\right)\left(\frac{12}{17}\right) = \left(\frac{2}{17}\right)^2\left(\frac{3}{17}\right) = \left(\frac{17}{3}\right) = \left(\frac{2}{3}\right) = -1,\end{aligned}$$

$$\begin{aligned}\left(\frac{83}{181}\right) &= \left(\frac{181}{83}\right) = \left(\frac{15}{83}\right) = \left(\frac{3}{83}\right)\left(\frac{5}{83}\right) = (-1)\left(\frac{83}{3}\right)\left(\frac{83}{5}\right)\\ &= (-1)\left(\frac{2}{3}\right)\left(\frac{3}{5}\right) = (-1)(-1)\left(\frac{5}{3}\right) = \left(\frac{2}{3}\right) = -1.\end{aligned}$$

3.27. If $n = 8r \pm 1$, then $\frac{n^2-1}{8} = 8r^2 \pm 2r$, which is even. If $n = 8r \pm 3$, then $\frac{n^2-1}{8} = 8r^2 \pm 6r + 1$, which is odd.

3.28. p is either 2 or 5 or is congruent to 1 or 4 modulo 5.

3.29. Obviously, there is no solution when $p = 2$, and there is a solution $x = 1$ then $p = 3$. Suppose now that $p > 3$. We can complete the square to get:

$$(2x-1)^2 \equiv -3 \mod p.$$

Hence, the congruence has solutions if and only if $\left(\frac{-3}{p}\right) = 1$. By the reciprocity law, we have

$$\left(\frac{-3}{p}\right) = \left(\frac{p}{3}\right) = \begin{cases}1 & p \equiv 1 \bmod 3\\ -1 & p \equiv 2 \bmod 3.\end{cases}$$

To summarize, the congruence has solutions if and only if $p = 3$ or $p \equiv 1 \bmod 3$.

3.30. If a is a primitive root, then $\left(\frac{a}{p}\right) = -1$. This implies $\left(\frac{-a}{p}\right) = \left(\frac{-1}{p}\right)\left(\frac{a}{p}\right) = (-1)(-1) = 1$, so $-a$ is not a primitive root modulo p.

3.31. The answers are 5, 2, 6 and 5 respectively. For example, in the case of the prime 193, we have $193 - 1 = 192 = 2^6 \times 3$. Therefore, the powers we need to check are $192/2 = 96$ and $192/3 = 64$.

$$2^{96} \equiv \left(\frac{2}{193}\right) = 1, \qquad 3^{96} \equiv \left(\frac{3}{193}\right) = \left(\frac{193}{3}\right) = 1,$$

$$5^{96} \equiv \left(\frac{5}{193}\right) = \left(\frac{193}{5}\right) = \left(\frac{3}{5}\right) = -1, \quad 5^{64} \equiv 625^{16} \equiv 46^{16} \equiv (-7)^8$$

$$\equiv 49^4 \equiv 85^2 \mod 193.$$

Since $85 \not\equiv \pm 1 \bmod 193$, it follows that $5^{64} \not\equiv 1 \bmod 193$. Therefore, 5 is the first primitive root modulo 193.

3.32. If $p \equiv \pm 1 \bmod 8$, then $\left(\frac{2}{p}\right) = 1$, so 2 is not a primitive root modulo p.

3.33. Note that $N!$ is a multiple of 8, and so p is congruent to 1 modulo 8, and also modulo every prime less than or equal to N. Choose any a between 1 and n. We can factorize a as $2^r q_1 \cdots q_s$ with odd primes $q_i \leq N$.

$$\left(\frac{a}{p}\right) = \left(\frac{2^r q_1 \cdots q_s}{p}\right) = \left(\frac{2}{p}\right)^r \left(\frac{q_1}{p}\right) \cdots \left(\frac{q_s}{p}\right).$$

Since $p \equiv 1 \bmod 8$, the Reciprocity Law and the Second Nebensatz imply that

$$\left(\frac{a}{p}\right) = 1^r \left(\frac{p}{q_1}\right) \cdots \left(\frac{p}{q_s}\right).$$

Since $p \equiv 1 \bmod q_i$, we have

$$\left(\frac{a}{p}\right) = \left(\frac{1}{q_1}\right) \cdots \left(\frac{1}{q_s}\right) = 1.$$

Therefore, a is a quadratic residue modulo p, so cannot be a primitive root modulo p.

3.34. Let $b^2 \equiv a \bmod p$. If $x^4 \equiv a \bmod p$, then $x^2 \equiv \pm b \bmod p$. Since $\left(\frac{-1}{p}\right) = -1$, either b or $-b$ is a quadratic residue, but not both. Suppose without loss of generality that b is a quadratic residue. Then we may take x to be one of the two square roots of b modulo p.

3.35. Since $\mathbb{F}_p^\times$ is a cyclic group whose order is a multiple of 4, there are exactly four elements y in $\mathbb{F}_p^\times$ such that $y^4 \equiv 1 \bmod p$. We'll call these elements $\pm 1, \pm i$. If x_0 is one solution to $x^4 \equiv a \bmod p$, then $\pm x, \pm ix$ are four solutions. Conversely, if x_1 is another solution, then clearly $(x_1/x_0)^4 = 1$ and so x_1 is one of the elements $\pm x, \pm ix$. This shows that if there is a solution to the congruence, then there are exactly 4 solutions.

Alternative proof: Since a is a quadratic residue, we have $a \equiv b^2 \bmod p$ for some b. Hence, we can factorize $x^4 - a$ to get

$$(x^2 + b)(x^2 - b) = 0.$$

Since $p \equiv 1 \bmod 4$, the First Nebensatz implies that $(-1/p) = 1$, and hence $(-b/p) = (b/p)$. This shows that $x^2 + b = 0$ has roots if and only if $x^2 - b$ has roots. Hence, there are either no solutions or 4 solutions.

Examples: Take $p = 5$. There are four solutions to $x^4 \equiv 1 \bmod 5$ but there are no solutions to $x^4 \equiv 4 \bmod 5$.

3.36. (1) If p is a factor of $4n^2 + 1$, then $(2n)^2 \equiv -1 \bmod p$. Hence, -1 is a quadratic residue modulo p. By the First Nebensatz, this implies $p \equiv 1 \bmod 4$.

(2) Assume that there are only finitely many prime numbers $p_1, \ldots, p_r$ which are congruent to 1 modulo 4. Let $N = 4(p_1 \cdots p_r)^2 + 1$ and let p be a prime factor of N. By the first part of the question, $p \equiv 1 \bmod 4$. However, p is not one of the p_i, since they are not factors of N. This gives a contradiction.

3.37. Suppose there are only finitely many such primes, and call them $p_1, \ldots, p_r$. Let $N = 4(p_1 \cdots p_r)^2 + 3$ and let q be any prime dividing of N. Then -3 is a quadratic residue modulo q. Using the Reciprocity Law, this implies that $q \equiv 1 \bmod 3$. Also, since q is odd, we know $q \equiv 1 \bmod 6$. Therefore, q is one of the primes p_i. This gives a contradiction since $N \equiv 3 \bmod p_i$.

3.38. If -2 and 3 are quadratic non-residues, then $\left(\frac{-6}{p}\right) = \left(\frac{-2}{p}\right)\left(\frac{3}{p}\right) = (-1)(-1) = 1$, so -6 is a quadratic residue. Therefore, at least one of the three is a quadratic residue. Suppose -2 is a quadratic residue, and write $\sqrt{-2}$ for a square root of -2 modulo p. Then we have

$$X^4 + 4X^2 + 1 \equiv (X^2 + \sqrt{-2}X + 1)(X^2 - \sqrt{-2}X + 1) \mod p.$$

Similarly in the other two cases, we have

$$X^4 + 4X^2 + 1 \equiv (X^2 + 2 + \sqrt{3})(X^2 + 2 - \sqrt{3}) \mod p,$$
$$X^4 + 4X^2 + 1 \equiv (X^2 + \sqrt{-6}X - 1)(X^2 - \sqrt{-6}X - 1) \mod p.$$

3.39. For odd integers n, m define $\chi(n, m) = (-1)^{\frac{(n-1)(m-1)}{4}}$. For odd integers n, n', we have

$$nn' - 1 \equiv (n-1) + (n'-1) \mod 4.$$

This implies $\chi(nn', m) = \chi(n, m)\chi(n', m)$, and since χ is symmetric in n and m we also have $\chi(n, mm') = \chi(n, m)\chi(nm')$. We'll use this fact to prove the Generalized Reciprocity Law. Let $n = p_1 \cdots p_r$ and let $m = q_1 \cdots q_s$. Then we have

$$\left(\frac{n}{m}\right) = \prod_{i,j} \left(\frac{p_i}{q_j}\right) = \prod_{i,j} \chi(p_i, q_j) \left(\frac{q_j}{p_i}\right) = \chi(n, m) \left(\frac{m}{n}\right).$$

To prove the generalized First Nebensatz, we must show that for odd n, m we have $(-1)^{\frac{nm-1}{2}} = (-1)^{\frac{n-1}{2}}(-1)^{\frac{m-1}{2}}$, and the Second Nebensatz we must show that $(-1)^{\frac{(nm)^2-1}{8}} = (-1)^{\frac{n^2-1}{8}}(-1)^{\frac{m^2-1}{8}}$. These follow from the congruences

$$nm - 1 \equiv (n-1) + (m-1) \mod 4, \quad (nm)^2 - 1 \equiv (n^2 - 1) + (m^2 - 1) \mod 16,$$

which are easy to prove for all odd integers n, m.

3.40. (1) Suppose there are only finitely many such primes, and call then $p_1, \cdots, p_r$, these primes are obviously not factors of n. By the Chinese remainder theorem, we can choose a positive integer N such that $\chi(N) = -1$ and $N \equiv 1 \mod p_1 \cdots p_r$. The prime factors q of N are not equal to any p_i, so we have $\chi(q) = 1$. Since N is a product of such primes, we must have $\chi(N) = 1$, which gives a contradiction.

(2) We may assume without loss of generality that n is not a multiple of any square apart from 1, so we are now trying to prove that $n = 1$. For a positive integer $x = p_1 \cdots p_r$ coprime to $2n$, define

$$\left(\frac{n}{x}\right) = \left(\frac{n}{p_1}\right) \cdots \left(\frac{n}{p_r}\right).$$

It's clear from the definition that $\left(\frac{n}{xy}\right) = \left(\frac{n}{x}\right)\left(\frac{n}{y}\right)$, and using the Reciprocity Law we can see that $\left(\frac{n}{x}\right)$ depends only on x modulo $4n$.

Therefore, we can interpret the function $x \mapsto \left(\frac{n}{x}\right)$ as a homomorphism $(\mathbb{Z}/4n)^\times \to \{1, -1\}$. We'll show that if $n \neq 1$ then this homomorphism is surjective. This is clear if $n = -1$. If $n \neq \pm 1$, then $n = qm$ with q a prime not dividing m. By the Chinese remainder theorem and the Reciprocity Law, we may choose an integer x such that $\left(\frac{q}{x}\right) = -1$ and $\left(\frac{m}{x}\right) = 1$. Hence, $\left(\frac{n}{x}\right) = -1$.

Now if $n \neq 1$ then by part (1) of this question, there are infinitely many primes p, such that $\left(\frac{n}{p}\right) = -1$. This gives a contradiction.

3.41. (*Reflexivity*) Since $x - x = 0 \cdot m$, we have $x \equiv x \bmod mR$.

(*Symmetry*) Suppose $x \equiv y \bmod mR$. This means $x - y = mz$ for some $z \in R$. Therefore, $y - x = m \cdot (-z)$ so $y \equiv x \bmod mR$.

(*Transitivity*) Suppose $x \equiv y \bmod mR$ and $y \equiv z \bmod mR$. This means $x - y = ma$ and $y - z = mb$ for elements $a, b \in R$. Therefore, $x - z = m(a+b)$, so $x \equiv z \bmod mR$.

3.42. (*Existence*) Choose polynomials h, k such that $1 = hf + kg$. If we set $c = hfb + kga$, then we have $c \equiv a \bmod f$ and $c \equiv b \bmod g$. This proves the existence of a solution.

(*Uniqueness*) Suppose that d is another solution to the congrunces. Then $c - d$ is a multiple of both f and g, and we must show that $d - c$ is a multiple of fg. We have $d - c = rf = sg$ for some polynomials r, s. From the equation $hf + kg = 1$, we get $d - c = rf = (hfr + kgr)f = (hsg + kgr)f = (hs + kr)fg$. Hence, $d - c$ is a multiple of fg.

3.43. We've already seen this in the example above when $p = 2$, so assume $p > 2$. If a is a quadratic non-residue modulo p, the the polynomial $f(X) = X^2 - a$ is irreducible modulo p, and therefore the quotient ring $\mathbb{F}_p[X]/f\mathbb{F}_p[X]$ is a field. The elements are $aX + b$ with $a, b \in \mathbb{F}_p$, so the field has p^2 elements.

3.44. We need to show that each of these numbers is a root of a monic polynomial with coefficients in $\mathbb{Z}$. Clearly, 5 is a roots of $X - 5$ and $\sqrt[3]{6}$ is a root of $X^3 - 6$. For $\sqrt{2} + \sqrt{3}$, note that $(\sqrt{2} + \sqrt{3})^2 = 5 + 2\sqrt{6}$. This implies $m(\sqrt{2} + \sqrt{3}) = 0$ where $m(X) = (X^2 - 5)^2 - 24$. Expanding out m, we get $m(X) = X^4 - 10X + 1$.

3.45. Lemma 3.4 shows that α^* is a bijection, so we only need to show that α^* is a ring homomorphism. To see this, note that $\alpha^*(f \times g) = (f \times g)(\alpha) = f(\alpha)g(\alpha) = \alpha^*(f)\alpha^*(g)$. A similar argument shows that $\alpha^*(f+g) = \alpha^*(f) + \alpha^*(g)$ and $\alpha^*(1) = 1$.

3.46. The elements are $0, 1, \sqrt{3}, 1+\sqrt{3}$. Note that $\sqrt{3}^2 \equiv 1 \bmod 2\mathbb{Z}[\sqrt{3}]$, so the multiplication table is:

	1	$\sqrt{3}$	$1+\sqrt{3}$
1	1	$\sqrt{3}$	$1+\sqrt{3}$
$\sqrt{3}$	$\sqrt{3}$	1	$1+\sqrt{3}$
$1+\sqrt{3}$	$1+\sqrt{3}$	$1+\sqrt{3}$	0

. This is not a field because $1+\sqrt{3}$ has no inverse.

3.47. Let p be an irreducible factor of f and let $\mathbb{E} = \mathbb{F}[X]/p\mathbb{F}[X]$. We know that $\mathbb{E}$ is a field because p is irreducible. Clearly, $p(X) \equiv 0 \bmod p(X)\mathbb{F}[X]$, and so the congruency class $X \bmod p\mathbb{F}[X]$ is a root of p (and hence of f) in the field $\mathbb{E}$.

3.48. Clearly, $F(1) \equiv 1 \bmod pR$ and $F(ab) \equiv F(a)F(b) \bmod pR$. The congruence $F(a+b) \equiv F(a) + F(b) \bmod pR$ follows from Lemma 3.5.

3.49. Assume first that R is a UFD. Suppose $ab \equiv 0 \bmod m$. This means that m is a factor of ab. Since R is a UFD and m is irreducible, it follows that m is a factor of a or of b. Hence, $a \equiv 0 \bmod m$ or $b \equiv 0 \bmod m$. This shows that R/mR is an integral domain.

Now assume that R is a Euclidean domain. If $a \not\equiv 0 \bmod m$, then since m is irreducible, it follows that a and m are coprime. Hence, there exist $h, k \in R$ such that $ha + km = 1$. This implies $ha \equiv 1 \bmod m$. We've shown that every non-zero element of R/mR has an inverse, so R/mR is a field.

3.50. Let $n(a)$ be the number of solutions to $x^2 \equiv a \bmod p$. With this notation, we have

$$\sum_{x=0}^{p-1} \zeta_p{}^{x^2} = \sum_{a=0}^{p-1} n(a)\zeta_p{}^a.$$

On the other hand, $n(0) = 1$ and $n(a) = \left(\frac{a}{p}\right) + 1$ for non-zero a, so we have

$$\sum_{x=0}^{p-1} \zeta_p{}^{x^2} = 1 + \sum_{a=1}^{p-1} \left(\left(\frac{a}{p}\right) + 1\right) \zeta_p{}^a = G(p) + \sum_{a=0}^{p-1} \zeta_p{}^a = G(p) + \Phi_p(\zeta_p) = G(p).$$

3.51. Let $f(X) = \sum_{a=1}^{p-1} \left(\frac{a}{p}\right) X^a$. Since $f(\zeta_p) = G(p)$, we know that $f(\zeta_p)^2 = p$ in the ring $\mathbb{Z}[\zeta]$. By Corollary 3.6, this implies $f^2 \equiv p \bmod \Phi_p\mathbb{Z}[X]$, so we have $f(X)^2 = q(X)\Phi_p(X) + p$ for some $q \in \mathbb{Z}[X]$.

The number $h = g^{\frac{q-1}{p}}$ has order p in $\mathbb{F}_q^\times$, so is a root of Φ_p in $\mathbb{F}_q$. Substituting h into f, we get $f(h)^2 = q(h)\Phi_p(h) + p \equiv p \bmod q$.

3.52. Let γ be a root of f in $\mathbb{Z}[\zeta_p]$. We have $\gamma = g(\zeta_p)$ for some $g \in \mathbb{Z}[X]$. This implies $f(g(\zeta_p)) = 0$, and hence by Corollary 3.6 $f(g(X)) \equiv 0 \bmod \Phi_p\mathbb{Z}[X]$. This is equivalent to $f(g(X)) = h(X)\Phi_p(X)$ for some $h \in \mathbb{Z}[X]$. Since $q \equiv 1 \bmod p$, the polynomial Φ_p has a root $a \in \mathbb{F}_q$ (a is an element of order p in $\mathbb{F}_q^\times$). Substituting a for X we get $f(g(a)) = h(a)\Phi_p(a) \equiv 0 \bmod q$, so $g(a)$ is a root of f in $\mathbb{F}_q$.

3.53. Let $\alpha = \zeta_7 + \zeta_7^{-1} = 2\cos(\frac{2\pi}{7})$. It follows that

$$\alpha^2 = \zeta_7^2 + 2 + \zeta_7^{-2}, \quad \alpha^3 = \zeta_7^3 + 3\zeta_7 + 3\zeta_7^{-1} + \zeta_7^{-3}.$$

Hence,

$$f(\alpha) = \zeta_7^3 + \zeta_7^2 + \zeta_7 + 1 + \zeta_7^{-1} + \zeta_7^{-2} + \zeta_7^{-3} = 0.$$

Clearly $\alpha \in \mathbb{Z}[\zeta_7]$ so the result follows by the previous exercise. Modulo 29 the roots of f are 3, 7 and 18.

3.54. If $p \equiv 1 \bmod 4$, then $G(p) = +\sqrt{p}$ and if $p \equiv 3 \bmod 4$, then $G(p) = +i\sqrt{p}$.

3.55. Choose a positive integer b such that $(a+1)(b+1) \equiv 1 \bmod p$. In particular, we have $\zeta_p{}^{(a+1)(b+1)} = \zeta_p$. Let $v = 1 + \zeta_p{}^{a+1} + \zeta_p{}^{2(a+1)} + \cdots + \zeta_p{}^{b(a+1)}$. We have (using the formula for a geometric progression)

$$uv = \frac{\zeta_p{}^{a+1} - 1}{\zeta_p - 1} \frac{\zeta_p{}^{(a+1)(b+1)} - 1}{\zeta_p{}^{a+1} - 1} = \frac{\zeta_p{}^{(a+1)(b+1)} - 1}{\zeta_p - 1} = \frac{\zeta_p - 1}{\zeta_p - 1} = 1.$$

Therefore, $u^{-1} = v \in \mathbb{Z}[\zeta_p]$.

4.1. The solutions are $x \equiv \pm 120 \bmod 7^4$, $x \equiv \pm 6397 \bmod 11^4$, $x \equiv \pm 3296 \bmod 11^4$, $x \equiv \pm 182 \bmod 5^4$, $x \equiv 162, 2238 \bmod 7^4$ and $x \equiv 212 \bmod 625$. The method is the same in each case. Here is the last one. We begin by noticing that $a_0 = 2$ is a solution modulo 5. Let $f(X) = X^3 - 3$. Then obviously $f'(X) = 3X^2$. We'll show that $a_0 = 2$ satisfies the conditions of Hensel's Lemma:

$$c = v_5(f'(a_0)) = v_5(12) = 0,$$
$$f(a_0) = 5 \equiv 0 \bmod 5^{2c+1}.$$

Define the sequence a_n by $a_{n+1} = a_n - \frac{a_n^3 - 3}{3a_n^2}$. By Hensel's Lemma, we know that $f(a_n) \equiv 0 \bmod 5^{2^n}$, so we need to calculate a_2 modulo 625.

$$a_1 = 2 - \frac{5}{12} \equiv 2 - 5 \cdot 3 \equiv 12 \mod 25,$$

$$a_2 \equiv 12 - \frac{1725}{432} \equiv 12 - \frac{25 \cdot (-6)}{7} \equiv 12 + 25 \cdot 8 \equiv 212 \mod 625.$$

The important point to notice here is that when reducing $\frac{1725}{432} = \frac{25 \cdot 69}{432}$ modulo 625, we need only calculate the congruency class of $\frac{69}{432}$ modulo 25.

4.2. If $x - y$ is a multiple of p^r, then $p^s x - p^s y$ is a multiple of p^{r+s}.

4.3. An element is a unit if and only if both x and x^{-1} are in $\mathbb{Z}_{(p)}$. This is equivalent to saying that $v_p(x) \geq 0$ and $-v_p(x) = v_p(x^{-1}) \geq 0$, i.e. $v_p(x) = 0$. This means that the reducible element are products xy where both x and y have valuation ≥ 1, so reducible elements have valuation ≥ 2. Conversely, if x has valuation at least 2 then $y = x/p$ has valuation at least 1, and x factorizes as $x = p \cdot y$. Therefore, x is reducible. It follows that the irreducible elements are those with valuation 1.

4.4. Factorizing n into primes, we have $n = p_1^{a_1} \cdots p_r^{a_r}$. This implies $v_{p_i}(n) = a_i$, which gives the first formula.

By uniqueness of factorization, the factors of n have the form $p_1^{b_1} \cdots p_r^{b_r}$ with $0 \leq b_i \leq a_i$. This means that $d|n$ if and only if $v_p(d) \leq v_p(n)$ for all prime numbers p. This implies d is a common factor of n and m if and only if $v_p(d) \leq v_p(n)$ and $v_p(d) \leq v_p(m)$. In particular, if d is the highest common factor, then $v_p(d) = \min(v_p(n), v_p(m))$. Similarly, if e is the lowest common multiple, then $v_p(e) = \max(v_p(n), v_p(m))$. We therefore have

$$\mathrm{hcf}(n, m) = \prod_p p^{\min(v_p(n), v_p(m))}, \quad \mathrm{lcm}(n, m) = \prod_p p^{\max(v_p(n), v_p(m))}.$$

Since $a+b = \min(a, b) + \max(a, b)$, we must have $nm = \mathrm{hcf}(n, m) \cdot \mathrm{lcm}(n, m)$.

4.5. If a is a root of Φ_7 modulo p^{1000}, then a is also a root modulo p. In particular, this implies $a^7 \equiv 1 \bmod p$, so a has order either 1 or 7 in $\mathbb{F}_p^\times$. If a has order 1, then $a \equiv 1 \bmod p$ so $0 \equiv \Phi_7(a) \equiv 7 \bmod p$, which is a contradiction since $p \neq 7$. Therefore, a has order 7 in $\mathbb{F}_p^\times$. By the corollary to Lagrange's Theorem, 7 is a factor of $p - 1$ so $p \equiv 1 \bmod 7$.

Conversely, assume that $p \equiv 1 \bmod 7$. Since 7 is a factor of $p - 1$, there is an element a_0 of order 7 in $\mathbb{F}_p^\times$. Such an element a_0 is a root of Φ_7, so we have

$$\Phi_7(a_0) \equiv 0 \mod p.$$

To show that a_0 satisfies the conditions of Hensel's Lemma, we need to show that $\Phi_7'(a_0) \not\equiv 0 \bmod p$. For this, we write $\Phi_7(X)$ as $\frac{X^7-1}{X-1}$ and differentiate:

$$\Phi_7'(X) = \frac{7X^6(X-1)-(X^7-1)}{(X-1)^2} = \frac{7X^6 - \Phi_7(X)}{X-1}.$$

This implies

$$(a_0-1)\Phi_7'(a_0) = 7a_0^6 - \Phi_7(a_0) \equiv 7a_0^6 \mod p.$$

Since $a_0 \not\equiv 0 \bmod p$ and $p \neq 7$, it follows that $\Phi_7'(a_0) \not\equiv 0 \bmod p$.

4.6. Let $f(x) = x^7+14x^5+28x^4+7x^3+49x^2+96x+62$. By Fermat's Little Theorem, the congruence $f(x) \equiv y \bmod 7$ is equivalent to $6x+6 \equiv y \bmod 7$, which has a solution because 6 is invertible modulo 7. Call this solution a_0. To prove that there is a solution modulo 7^n, we need to show that a_0 satisfies the conditions of Hensel's Lemma. We have $f'(a_0) \equiv 5 \bmod 7$, so $v_7(f'(a_0)) = 0$. This shows the existence of a solution. We've shown that the polynomial function $f : \mathbb{Z}/p^n \to \mathbb{Z}/p^n$ is surjective. This function must also be injective, since $\mathbb{Z}/p^n$ is finite. Therefore, solutions are unique.

4.7. (1) Suppose a is an approximate root of g. We have $g'(X) = 2f(X)f'(X)$. Therefore,

$$c = v_p(g'(a)) = v_p(2f'(a)) + v_p(f(a)) \geq v_p(f(a)).$$

Since a is an approximate root, we have $v_p(g(a)) \geq 2c+1$. Since $g = f^2$, this implies

$$2v_p(f(a)) \geq 2c+1 \geq 2v_p(f(a)) + 1.$$

This is a contradiction.

(2) We'll first show that f and f' are coprime in $\mathbb{Q}[X]$. Suppose they have a common factor $g(X)$. Then any complex root of g is a common root of f and f', so is a double root of f. We're assuming f has no repeated roots, so f and f' are coprime. This implies that there are polynomials $h_0(X), k_0(X) \in \mathbb{Q}[X]$ such that $h_0 f + k_0 f' = 1$. It's possible that h_0 and k_0 don't have coefficients in $\mathbb{Z}_{(p)}$, but we can always find a power p^r such that $h = p^r h_0$ and $k = p^r k_0$ have coefficients in $\mathbb{Z}_{(p)}$. Therefore, $hf + kf' = p^r$.

Let a_0 be a root of f modulo p^{2r+1}. Then we have

$$0 \not\equiv p^r = h(a_0)f(a_0) + k(a_0)f'(a_0) \equiv k(a_0)f'(a_0) \mod p^{r+1}.$$

Therefore, $c = v_p(f'(a_0)) \leq r$ so $v_p(f(a_0)) \geq 2r+1 \geq 2c+1$. This shows that a_0 is an approximate root.

4.8. (1) $x = 1$ is a solution.

(2) There are no solutions because there are none modulo 4.

(3) There are solutions because $(\frac{87}{127}) = 1$.

(4) There are solutions by the Chinese remainder theorem ($x = 0$ is a solution modulo 3). To get a solution module 127^{2000}, we use Hensel's Lemma.

(5) There are no solutions, because there are none modulo 9.

(6) There are no solutions. Suppose x is a solution. Then x is a multiple of 29. But then x^2 must be a multiple of 29^2, which contradicts $x^2 \equiv 83 \bmod 29^2$.

4.9. We first answer the question in the case that n is a power of a prime. We have the solution 0 modulo 5 but there is no solution modulo 25. Furthermore, 1 is a solution modulo 2 and modulo 4, but there are no solutions modulo 8. Suppose now that p is a prime which is not equal to 2 or 5. Using the Reciprocity Law, we have

$$\left(\frac{5}{p}\right) = \left(\frac{p}{5}\right) = \begin{cases} 1 & p \equiv \pm 1 \bmod 5, \\ -1 & p \equiv \pm 2 \bmod 5. \end{cases}$$

Hence, there are solutions modulo p^a when $p \equiv \pm 1 \bmod 5$ but not when $p \equiv \pm 2 \bmod 5$.

Now using the Chinese remainder theorem, we see that there are solutions modulo n if and only if $n = n_1 n_2$, where n_1 is a factor of 20 and n_2 is a product of primes congruent to ± 1 modulo 5 to any powers.

4.10. Assume p is an odd prime. Then by completing the square, where are solutions modulo p^m if and only if $y^2 \equiv 17 \bmod p^m$ has solutions. If p is neither 2 nor 17, then this only depends on $\left(\frac{17}{p}\right)$. By the Quadratic Reciprocity Law, we can show that $\left(\frac{17}{p}\right) = 1$ if and only if p is congruent to ± 1, ± 2, ± 4 or ± 7 modulo 17. Hence, there are solutions modulo p^m for those primes.

Modulo 17 there is a solution, but there are no solutions modulo 17^2, and therefore there are no solutions modulo any higher powers of 17.

When $p = 2$, we cannot complete the square, so we use Hensel's Lemma directly on $f(X) = X^2 + X - 4$. We have $f'(X) = 2X + 1$. Therefore, $f(1) \equiv 0 \bmod 2$ and $v_2(f'(1)) = 0$, so a_0 satisfies the conditions of Hensel's Lemma with the prime 2. Therefore, there are solutions modulo all powers of 2.

Putting this together, there are solutions modulo n if and only if $n = ab$, where a is either 1 or 17 and b is a product of primes congruent to ± 1, ± 2, ± 4 or ± 7 modulo 17, raised to arbitrarily high powers.

4.11. The number 0 is a solution modulo p. Suppose x is a solution modulo p^2. Then $x^2 - p$ is a multiple of p^2, and therefore x^2 is a multiple of p, which implies that x is a multiple of p. However, this implies that x^2 is a multiple of p^2, so we actually have $-p \equiv 0 \bmod p^2$. This is a contradiction.

4.12. By the Chinese Remainder Theorem, it's sufficient to check this modulo all prime powers p^n.

Assume first that p is a prime different from 2 and 7. We have

$$\left(\frac{2}{p}\right)\left(\frac{-7}{p}\right)\left(\frac{-14}{p}\right) = \left(\frac{-14}{p}\right)^2 = 1.$$

Hence not all three of these quadratic residue symbols are -1, so at least one of them is 1. Suppose without loss of generality that $\left(\frac{2}{p}\right) = 1$. Then the congruence $x^2 - 2 \equiv 0 \bmod p$ has solutions, and these can be lifted by Hensel's Lemma to solutions modulo p^n. If $x^2 - 2 \equiv 0 \bmod p^n$, then clearly $f(x) \equiv 0 \bmod p^n$.

Next, consider the prime 2. Since $-7 \equiv 1 \bmod 8$, it follows that $x^2 \equiv -7 \bmod 2^n$ has solutions for all n.

Finally, consider the prime 7. We have $\left(\frac{2}{7}\right) = 1$. So $x^2 - 2 \equiv 0 \bmod 7^n$ has solutions.

Hence, $f(x) \equiv 0 \bmod n$ has solutions for all n.

4.13. (1) $\Longrightarrow$ (2) Let $a = 2^r b$ with b odd and let $x^2 \equiv a \bmod 2^{r+3}$. Then $r = v_2(a) = 2v_2(x)$. In particular, r is even and $x = 2^{r/2}y$. But then $y^2 \equiv b \bmod 8$. This implies $b \equiv 1 \bmod 8$.

(2) $\Longrightarrow$ (1) Suppose $a = 4^m b$, where $b \equiv 1 \bmod 8$. We've already seen that b is a square modulo 2^n. Therefore, a is also a square modulo 2^n.

4.14. Let x_0 be a solution to $f(x) \equiv 0 \bmod p$ where $f(X) = X^3 - a$. Since $a \not\equiv 0 \bmod p$, it follows that $x_0 \not\equiv 0 \bmod p$. Therefore, $f'(x_0) = 3x_0^2 \not\equiv 0 \bmod p$. Therefore, x_0 satisfies the conditions of Hensel's Lemma.

4.15. By Lemma 4.4, we have

$$\begin{aligned} v_2(120!) &= \left\lfloor \frac{120}{2} \right\rfloor + \left\lfloor \frac{120}{4} \right\rfloor + \left\lfloor \frac{120}{8} \right\rfloor + \left\lfloor \frac{120}{16} \right\rfloor + \left\lfloor \frac{120}{32} \right\rfloor + \left\lfloor \frac{120}{64} \right\rfloor \\ &= 60 + 30 + 15 + 7 + 3 + 1 = 116. \end{aligned}$$

$$\begin{aligned} v_3\left(\frac{100!}{20!80!}\right) &= \left\lfloor\frac{100}{3}\right\rfloor + \left\lfloor\frac{100}{9}\right\rfloor + \left\lfloor\frac{100}{27}\right\rfloor + \left\lfloor\frac{100}{81}\right\rfloor \\ &\quad - \left(\left\lfloor\frac{20}{3}\right\rfloor + \left\lfloor\frac{20}{9}\right\rfloor\right) - \left(\left\lfloor\frac{80}{3}\right\rfloor + \left\lfloor\frac{80}{9}\right\rfloor + \left\lfloor\frac{80}{27}\right\rfloor\right) \\ &= 33 + 11 + 3 + 1 - 6 - 2 - 26 - 8 - 2 = 4. \end{aligned}$$

$$v_{11}((11^5)!) = 11^4 + 11^3 + 11^2 + 11 + 1 = \frac{11^5 - 1}{11 - 1} = 16105.$$

4.16. (1) $v_p(p^n) = n \to \infty$, so the series converges p-adically.

(2) $v_p(p^n + 1) = 0$, so the series does not converges p-adically.

(3) $v_p(n!) \geq \lfloor \frac{n}{p} \rfloor \to \infty$, so this converges.

(4) $v_p(n)$ is zero for infinitely many n (for all n which are not multiples of p), so this does not converge.

(5) By the binomial theorem, $\frac{(2n)!}{n!}$ is a multiple of $n!$. This implies $v_p(\frac{(2n)!}{n!}) \geq v_p(n!) \to \infty$. Therefore, the series converges p-adically.

(6) This series is a little tricky, because the valuations are unbounded, but they do not converge to infinity because there are infinitely many terms whose valuation is 0 or 1. Therefore, the series does not converge. To see why this is true, note that $v_p(n!) = \lfloor \frac{n}{p} \rfloor + \lfloor \frac{n}{p^2} \rfloor + \ldots$. This implies $v_p((p^a)!) = 1 + p + p^2 + \cdots + p^{a-1}$. If we assume $p > 2$, then we have $v_p((2p^a)!) = 2 + 2p + 2p^2 + \cdots + 2p^{a-1} = 2v_p((p^a)!)$. This implies $v_p\left(\frac{(2p^a)!}{(p^a)!^2}\right) = 0$, so there are infinitely many terms with valuation 0.

In the case $p = 2$, a similar argument shows that $v_2\left(\frac{(2\cdot 2^a)!}{(2^a)!^2}\right) = 1$. Hence, there are infinitely many terms with valuation 1 in this case.

4.17. Choose any natural number r. Since $\sum a_n$ and $\sum b_n$ converge p-adically, there is an N such that for all $n > N$ we have $a_n \equiv b_n \equiv 0 \bmod p^r$. Hence, $a_n + b_n \equiv 0 \bmod p^r$. If $n > 2N$, then this implies $c_n \equiv 0 \bmod p^r$. This shows that the series $\sum(a_n + b_n)$ and $\sum c_n$ both converge p-adically. The first congruence follows because

$$\sum_{n=1}^{N}(a_n + b_n) \equiv \sum_{n=1}^{N} a_n + \sum_{n=1}^{N} b_n \mod p^r.$$

The second congruence follows because

$$\sum_{n=0}^{2N} c_n = \sum_{i+j \leq 2N} a_i b_j \equiv \sum_{i \leq N} \sum_{j \leq N} a_i b_j \equiv \left(\sum_{i \leq N} a_i\right)\left(\sum_{j \leq N} b_j\right) \mod p^r.$$

4.18. For $n \geq r$, we have $(px)^n \equiv 0 \bmod p^r$. Therefore, using the formula for a geometric progression, we have:

$$\sum_{n=0}^{\infty}(px)^n \equiv \sum_{n=0}^{r-1}(px)^n \equiv \frac{(px)^r - 1}{px - 1} \mod p^r.$$

Since the denominator $px - 1$ is invertible modulo p^r, we may reduce the numerator modulo p^r. This gives us

$$\sum_{n=0}^{\infty}(px)^n \equiv \frac{-1}{px-1} \equiv \frac{1}{1-px} \mod p^r.$$

Substituting $p = 5$ and $x = -2$, we find $11^{-1} \equiv 1 - 10 + 100 - 1000 + \cdots \equiv 341 \bmod 625$.

4.19. By Lemma 4.4, the left-hand side of the inequality is equal to $\sum_{a=1}^{\infty}(\frac{n}{p^a} - \lfloor \frac{n}{p^a} \rfloor)$. We can divide this into two sums as follows:

$$\sum_{1 \leq a \leq \log_p(n)} \left(\frac{n}{p^a} - \left\lfloor \frac{n}{p^a} \right\rfloor\right) + \sum_{a > \log_p(n)} \frac{n}{p^a}.$$

In the first of these sums, each term is less than 1, and there are $\leq \log_p(n)$ terms, so the total is $\leq \log_p(n)$. The second sum is a geometric progression with first term < 1 and ratio of terms $\frac{1}{p}$. Therefore, the second sum is $< \frac{p}{p-1}$.

4.20. Let F_n be the coefficient of X^n in the power series $f(X + X^2)$; we'll show that F_n is the Fibonacci sequence. It's easy to check that $F_0 = F_1 = 1$, so we just need to check the recurrence relation. We have $(1-(X+X^2))f(X+X^2) = \sum(X+X^2)^n - \sum(X+X^2)^{n+1} = 1$. For $n \geq 0$, the coefficient of X^{n+2} in $(1 - X - X^2))f(X + X^2)$ is $F_{n+2} - F_{n+1} - F_n$, so we have $F_{n+2} = F_{n+1} + F_n$.

4.21.

$$\begin{aligned}
\log(1+5x) &\equiv 5x - \frac{5^2}{2}x^2 + \frac{5^3}{3}x^3 \mod 5^4, \\
&\equiv 5x + 5^2 \cdot 12x^2 + 5^3 \cdot 2x^3 \mod 5^4, \\
\exp(5x) &\equiv 1 + 5x + \frac{5^2}{2}x^2 + \frac{5^3}{6}x^3 \mod 5^4, \\
&\equiv 1 + 5x + 5^2 \cdot 13x^2 + 5^3x^3 \mod 5^4.
\end{aligned}$$

Using the formula for $\log(1+5x)$, we get $\log(11) \equiv \log(1+5\cdot 2) \equiv 85 \bmod 5^4$. Therefore,

$$11^a \equiv \exp(a\log(11)) \equiv 1 + 5 \cdot 17a + 5^2 \cdot 7a^2 + 5^3 \cdot 3a^3 \mod 625.$$

To solve the congruence, we take $x \equiv 11^{\frac{1}{4}} \bmod 625$. This gives

$$x \equiv 1 + \frac{5 \cdot 17}{4} + \frac{5^2 \cdot 7}{16} + \frac{5^3 \cdot 3}{64} \equiv 166 \mod 625.$$

4.22. The formula for the logarithm is $\log(1+3x) \equiv 3x+36x^2+9x^3 \bmod 81$. Substituting $x = -2$ and $x = 5$, we get

$$\log(-5) \equiv 66 \bmod 81, \quad \log(16) \equiv 15 \bmod 81.$$

Suppose $16^y \equiv -5 \bmod 81$. Then we have $y \log(16) \equiv \log(-5) \bmod 81$. By (b), we have $15y \equiv 66 \bmod 81$. Therefore, $y \equiv -1 \bmod 27$. Clearly, $x = 4y$ is a solution, so we have $x \equiv -4 \equiv 50 \bmod 54$. (Note also, $\log(2)$ does not converge, so any argument involving $\log(2)$ is wrong.)

4.23. By definition, $\exp(2) = \sum \frac{2^n}{n!}$. The valuation of the n-th term is $n - v_2(n!)$. If $n = 2^a$, then

$$v_2(n!) = 2^{a-1} + 2^{a-2} + \cdots + 2 + 1 = 2^a - 1 = n - 1.$$

Therefore, such terms have valuation 1. Hence, the valuation does not tend to infinity, so the series does not converge 2-adically.

The n-th term of $\exp(4x)$ is $\frac{4^n x^n}{n!}$ and this has valuation

$$v_2\left(\frac{4^n x^n}{n!}\right) \geq 2n - \frac{n}{2-1} = n \to \infty.$$

Therefore, $\exp(4x)$ converges 2-adically.

4.24. The nth term of $\log(1+2x)$ is $\pm\frac{2^n x^n}{n}$, and this has valuation at least $n - \frac{\log_{\mathbb{R}}(n)}{\log_{\mathbb{R}}(2)}$, which tends to infinity. Therefore, the series converges 2-adically. Let $g(x) = \log(1+2x)$. For small real x, we have

$$g(f(x)) = \log(1 + 4x + 4x^2) = \log((1+2x)^2) = 2\log(1+2x) = 2g(x).$$

Hence, by the power series trick, we have the following congruence for all N and all $x \in \mathbb{Z}_{(2)}$:

$$\log(1 + 2f(x)) \equiv 2\log(1+2x) \mod 2^N.$$

Substituting $x = -1$, we get

$$\log(1) \equiv 2\log(1-2) \mod 2^N.$$

Obviously $\log(1) = 0$, so expanding $\log(1-2)$ as a series we get

$$2\left(-2 - \frac{2^2}{2} - \frac{2^3}{3} - \cdots\right) \equiv 0 \mod 2^N.$$

Cancelling this, we get

$$\sum_{n=1}^{\infty} \frac{2^n}{n} \equiv 0 \mod 2^{N-1}.$$

To prove the equation in (4), note that the derivative of $\mathrm{Li}(x)$ is $-\frac{\log(1-x)}{x}$. Using this, we can show (where all terms converge, i.e. for small real numbers x) that

$$\frac{d}{dx}\left(\mathrm{Li}_2(x) + \mathrm{Li}_2\left(\frac{x}{x-1}\right) + \frac{1}{2}\log(1-x)^2\right) = 0.$$

Hence, the function being differentiated is a constant in the region of convergence. Substituting $x = 0$, we see that the constant is 0. We can rewrite this equation as $f(g(x)) = h(x)$, where $f(x) = 2\mathrm{Li}_2(2x)$, $g(x) = \frac{x}{2x-1} = -x - 2x^2 - 4x^3 - \cdots$ and $h(x) = -\log(1-2x)^2 - 2\mathrm{Li}_2(2x)$. Using the power series trick and substituting $x = 1$, we get

$$4\mathrm{Li}_2(2) \equiv -\log(-1)^2 \mod 2^N.$$

We've already shown that $\log(-1) \equiv 0 \bmod 2^N$, so we have $\mathrm{Li}_2(2) \equiv 0$ modulo all powers of 2.

4.25. (1) Recall that $\exp : p\mathbb{Z}/p^n \to 1 + p\mathbb{Z}/p^n$, so its image is congruent to 1 modulo p.

(2) $\exp(\log(a)) = a$, since exp and log are inverse functions.

(3) Since exp and log are group homomorphisms between the additive and multiplicative groups, $\log(ab) = \log(a) + \log(b)$ and therefore

$$(ab)^x \equiv \exp(x\log(a) + x\log(b)) \equiv \exp(x\log(a))\exp(x\log(b)) \equiv a^x b^x.$$

(4) Since exp is a group homomorphism, we have

$$\exp((x+y)\log(a)) \equiv \exp(x\log(a) + y\log(a)) \equiv \exp(x\log(a))\exp(y\log(a)).$$

(5) Since log and exp are inverse functions, we have

$$\log(a^x) \equiv \log(\exp(x\log(a))) \equiv x\log(a) \mod p^n.$$

Therefore,

$$(a^x)^y \equiv \exp(yx\log(a)) \equiv a^{xy} \mod p^n.$$

4.26. The terms in the power series $(1+px)^a$ are $\frac{a(a-1)\cdots(a-n+1)}{n!}(px)^n$. Since x and a are in $\mathbb{Z}_{(p)}$, we have $v_p(a(a-1)\cdots(a-n+1)x^n) \geq 0$. Therefore, the valuation of this term is at least $v_p(\frac{p^n}{n!}) \geq \frac{p-2}{p-1}n$. Since $p > 2$, this tends to infinity, so the series converges p-adically.

Let $f(X) = \exp(X)$, $g(X) = a\cdot\log(1+pX)$ and let $h(X)$ be the binomial expansion of $(1+pX)^a$. If $|x| < \frac{1}{p}$, then we have $f(g(x)) = h(x)$. Therefore, by The Power Series Trick, $h = f \circ g$ as power series.

4.27. $T(1) = 1$, and $T(2) = T(-1) = (-1)^{3^{99}} = -1$.

4.28. (1) We saw already that $2^{25} \equiv 57 \bmod 125$. This implies $2^{125} \equiv 57^5 \bmod 625$. We can calculate this by the binomial expansion

$$2^{125} \equiv (7+50)^5 \equiv 7^5 + 5\cdot 7^4 \cdot 50 \mod 625.$$

Since $7^4 \equiv 1 \bmod 5$, we have $5\cdot 7^4\cdot 50 \equiv 250 \bmod 625$. We can also expand out 7^5 using the binomial theorem:

$$(2+5)^5 \equiv 2^5 + 5\cdot 2^4\cdot 5 + 10\cdot 2^3\cdot 25 \equiv 32 + 400 + 125 \equiv 557 \mod 625.$$

This gives $T(2) \equiv 182 \bmod 625$. We also have $T(1) = 1$, $T(4) = -1$ and $T(3) = -T(2) \equiv 443 \bmod 625$.

(2) Note that $13 \equiv 3 \bmod 5$, so we have $x = 3$. Hence,

$$5y \equiv \log(13T(3)^{-1}) \equiv \log(13T(2)) \equiv \log(13\cdot 182) \equiv \log(491) \mod 625.$$

This gives

$$5b \equiv \log(1+490) \equiv 490 - \frac{490^2}{2} + \frac{490^3}{3} \equiv 315 \bmod 625.$$

(3) We have

$$13^{234} \equiv T(3^{234})\cdot\exp(234\cdot 315).$$

By Fermat's little theorem, $3^4 \equiv 1 \bmod 5$, so $3^{234} \equiv 9 \equiv 4 \bmod 5$. Therefore,

$$\begin{aligned} 13^{234} &\equiv T(4)\cdot\exp(5\cdot 117) && \mod 625 \\ &\equiv -\left(1 + 5\cdot 117 + \frac{25\cdot 117^2}{2} + \frac{125\cdot 117^3}{6}\right) \equiv 114 && \mod 625. \end{aligned}$$

(4) The solution is $T(3^{47^{-1}})\cdot\exp(5\cdot\frac{63}{47})$. By Fermat's little theorem, $3^4 \equiv 1 \bmod 5$, so we have $3^{47^{-1}} \equiv 3^{-1} \equiv 2 \bmod 5$. Also, $63/47 \equiv 4 \bmod 125$. Therefore,

$$x \equiv T(2)\exp(5\cdot 4) \equiv 182\cdot\left(1 + 20 + \frac{400}{2} + \frac{8000}{6}\right) \equiv 597 \mod 625.$$

4.29. (1) We have $2^{11} = 2048 \equiv 2 - 11 \bmod 11$. Therefore,

$$\begin{aligned} 2^{11^2} &\equiv (2-11)^{11} \equiv 2^{11} - 11\cdot 2^{10}\cdot 11 && \mod 11^3, \\ &\equiv 2048 - 11^2 \equiv 596 && \mod 11^3. \end{aligned}$$

Therefore, $T(2) \equiv 596 \bmod 11^3$.

(2) The number 2 is a primitive root modulo 11, so every element of $\mathbb{F}_{11}^{\times}$ can be written in the form 2^c. Since T is a homomorphism, we have $T(2^c) = T(2)^c \equiv 596^c \bmod 1331$. Using this idea, we get

$$T(4) \equiv 596^2 \equiv 355216 \equiv 1170, \qquad T(8) \equiv 596 \cdot 1170 \equiv 697320 \equiv 1207,$$

$$T(5) \equiv 596 \cdot 1207 \equiv 632, \qquad T(10) \equiv -1 \equiv 1330,$$

$$T(9) \equiv -T(2) \equiv 735, \qquad T(7) \equiv -T(4) \equiv 161,$$

$$T(6) \equiv -T(5) \equiv 699, \qquad T(3) \equiv -T(8) \equiv 124.$$

(3) The solutions are:

$$6 \equiv T(6)\exp(11 \cdot 6), \qquad 17 \equiv T(6)\exp(11 \cdot 85),$$

$$35 \equiv T(2)\exp(11 \cdot 24), \qquad 47 \equiv T(3)\exp(11 \cdot 82).$$

For example, here is the calculation in the case of the number 35. Since $35 \equiv 2 \bmod 11$, we have $a \equiv 2 \bmod 11$. Therefore,

$$\exp(11b) \equiv 35 \cdot T(2^{-1}) \equiv 35 \cdot T(6) \equiv 507 \quad \bmod 1331.$$

Taking logarithms of this, we get

$$11b \equiv \log(1 + 506) \equiv 506 - 506^2/2 \equiv 264 = 11 \cdot 24 \quad \bmod 1331.$$

4.30. (1) Letting $x = T(a)\exp(7b)$, we get

$$T(a^4)\exp(28b) \equiv T(1)\exp(112) \quad \bmod 7^{30}.$$

We, therefore, have

$$a^4 \equiv 1 \bmod 7, \quad 28b \equiv 112 \bmod 7^{30}.$$

Note that $a^2 \equiv \pm 1 \bmod 7$, but by the First Nebensatz, -1 is not a square modulo 7 so in fact $a^2 \equiv 1 \bmod 7$. Therefore, $a = \pm 1 \bmod 7$.

The second congruence has solutions $7b \equiv 28 \bmod 7^{30}$. We, therefore, have two solutions to the congruence

$$x \equiv T(1)\exp(28), \quad T(-1)\exp(28) \quad \bmod 7^{30}.$$

(2) Doing the same as above with the congruence $x^7 \equiv T(3)\exp(98)$, we get

$$T(a^7)\exp(49b) \equiv T(3)\exp(98) \quad \bmod 7^{30}.$$

Hence, $a^7 \equiv 3 \bmod 7$ and $49b \equiv 98 \bmod 7^{30}$. By Fermat's little theorem, we know that $a^7 \equiv a \bmod 7$. Therefore, $a \equiv 3 \bmod 7$. We also have $7b \equiv 14 \bmod 7^{29}$. This means that modulo 7^{30} we have

$$7b \equiv 2 + 7^{29}c \quad \bmod 7^{30}, \quad c = 0, 1, 2, 3, 4, 5, 6.$$

We therefore have 7 solutions:

$$x \equiv T(3)\exp(14 + 7^{29}c) \mod 7^{30}, \quad c = 0, 1, 2, 3, 4, 5, 6.$$

Equivalently, $x \equiv T(3)\exp(14) \bmod 7^{29}$.

4.31. Let $x = T(a)\exp(pb)$. Then we have $T(a^8)\exp(8pb) \equiv 1 \bmod p^n$. Therefore, $a^8 \equiv 1 \bmod p$ and $8pb \equiv 0 \bmod p^n$. The second congruence shows (since p is odd) that $pb \equiv 0 \bmod p^n$, so in particular $x = T(a)$. The congruence $a^8 - 1 \equiv 0 \bmod p$ factorizes as

$$(a^4 + 1)(a^2 + 1)(a + 1)(a - 1) \equiv 0 \mod p.$$

Since $p \equiv 3 \bmod 4$, the First Nebensatz tells us that -1 is a quadratic non-residue modulo p, so $a^4 + 1$ and $a^2 + 1$ are invertible modulo p. From this, we see that $a = \pm 1$. Hence, the solutions are $x = T(1) = 1$ and $x = T(-1) = -1$.

4.32. (1) Let $x = T(g)\exp(p)$ and suppose $x^d \equiv 1 \bmod p^n$. Then $T(g^d)\exp(dp) \equiv T(1)\exp(0) \bmod p^n$. This implies

$$g^d \equiv 1 \bmod p, \quad dp \equiv 0 \bmod p^n.$$

Since g has order $p - 1$ modulo p, we have

$$d \equiv 0 \bmod p - 1 \quad \text{and} \quad d \equiv 0 \bmod p^{n-1}.$$

By the Chinese remainder theorem, this implies $d \equiv 0 \bmod (p-1)p^{n-1}$. Hence, the order of x is $(p-1)p^{n-1} = \varphi(p^n)$, so x generates the whole group.

(2) The equation $(x, y)^n = (1, 1)$ is equivalent to $x \equiv 0 \bmod \mathrm{ord}(x)$ and $x \equiv 0 \bmod \mathrm{ord}(y)$. Therefore, $\mathrm{ord}(x, y) = \mathrm{lcm}(\mathrm{ord}(x), \mathrm{ord}(y))$. In particular, $\mathrm{ord}(x, y) = |G| \cdot |H|$ if and only if $\mathrm{ord}(x) = |G|$, $\mathrm{ord}(y) = |H|$ and $\mathrm{hcf}(|G|, |H|) = 1$.

(3) The multiplicative group is cyclic if and only if $n = p^a$ or $n = 2p^a$ or $n = 4$, where p is an odd prime and $a \geq 0$ is an integer. This follows from the second part of the question, together with the decomposition of $(\mathbb{Z}/n)^\times$ as a product of cyclic groups.

4.33. By Fermat's Little Theorem, we have $f(a_0) \equiv 0 \bmod p$. Furthermore, $f'(a_0) = (p-1)a_0^{p-2} \not\equiv 0 \bmod p$ so a_0 satisfies Hensel's Lemma. Since a_n is a root of f modulo p^{2^n}, it follows that $a_n^{p-1} \equiv 1 \bmod p^{2^n}$. If we decompose a_n as $T(x)\exp(py)$, then this equation becomes $(p-1)py \equiv 0 \bmod p^{2^n}$. Solving this, we get $py \equiv 0 \bmod p^{2^n}$ so $a_n \equiv T(x) \bmod p^{2^n}$ for some x. On the other hand, $T(x) \equiv x \bmod p$ and $a_n \equiv a_0 \bmod p$, so we have $x \equiv a_0 \bmod p$. This implies $a_n = T(a_0) \bmod p^{2^n}$.

4.34. (1) Note that all terms in $\log(1+4x)$ are multiples of 4, and all but the constant term in $\exp(4x)$ are multiples of 4. By the power series trick, it follows that

$$\exp\left(4 \cdot \frac{\log(1+4x)}{4}\right) \equiv 1 + 4x \mod 2^n,$$
$$\log\left(1 + 4\left(\frac{\exp(4x)-1}{4}\right)\right) \equiv 4x \mod 2^n.$$

We therefore have a bijection $\exp : 4\mathbb{Z}/2^n \to 1 + 4\mathbb{Z}/2^n$, whose inverse is log.

If $x \in (\mathbb{Z}/2^n)^\times$, then $x \equiv \pm 1 \bmod 4$. Hence, $x \equiv \pm y \bmod 2^n$ for a unique $y \in 1 + 4\mathbb{Z}/2^n$. This implies $x \equiv \pm \exp(4z) \bmod 2^n$ for a unique $z \in 4\mathbb{Z}/2^n$.

(2) Suppose $x^2 \equiv 1 \bmod 2^n$, and let $x \equiv \pm \exp(4b) \bmod 2^n$. This implies $8x \equiv 0 \bmod 2^n$, so $4n$ is congruent to either 0 or 2^{n-1} modulo 2^n. Hence, $\exp(4n)$ is either 1 or $1 + 2^{n-1}$ modulo 2^n. This shows that the four roots of $x^2 - 1$ modulo 2^n are 1, -1, $1 + 2^{n-1}$ and $-1 - 2^{n-1}$. The polynomial factorizes in two ways as follows:

$$X^2 - 1 \equiv (X+1)(X-1) \equiv (X + 2^{n-1} + 1)(X - 2^{n-1} - 1) \mod 2^n.$$

4.35. Let $g(\zeta_p)$ be a root of f, where $g \in \mathbb{Z}[X]$. The equation $f(g(\zeta_p)) = 0$ is equivalent to the congruence $f(g(X)) \equiv 0 \bmod \Phi_p\mathbb{Z}[X]$. If $q \equiv 1 \bmod p$, then $\mathbb{F}_q^\times$ has an element y of order p. This implies $y^p \equiv 1 \bmod p$ and $y \not\equiv 1 \bmod p$. Therefore, $z^p \equiv 1 \bmod q^n$ where $z \equiv T(y) \bmod q^n$. Since $z \equiv y \not\equiv 1 \bmod q$, it follows that $z - 1$ is invertible modulo q^n. Therefore, $\Phi_p(z) = (z^p - 1)/(z-1) \equiv 0 \bmod q^n$. We've shown that $f(g(X))$ is a multiple of $\Phi_p(X)$. This implies $f(g(z))$ is a multiple of $\Phi_p(z)$, which is a multiple of q^n. Therefore, $g(z)$ is a root of f in $\mathbb{Z}/q^n$.

4.36. This follows because $\max\{|x|_p, |y|_p\} \leq |x|_p + |y|_p$.

4.37. (Reflexivity) Since $x_n \equiv x_n \bmod p^r$, it follows that x_n is equivalent to x_n, so equivalence of Cauchy sequences is transitive. (Symmetry) If $x_n \equiv y_n \bmod p^r$, then $y_n \equiv x_n \bmod p^r$, so equivalence of Cauchy sequences is transitive. (Transitivity) Suppose x_n is equivalent to y_n and y_n is equivalent to z_n. This means that for any r there exist N_1 and N_2, such that if $n > N_1$ then $x_n \equiv y_n \bmod p^r$ and if $n > N_2$ then $y_n \equiv z_n \bmod p^r$. If n is bigger than $\max\{N_1, N_2\}$, then both of these congruences hold, and therefore $x_n \equiv z_n \bmod p^r$. Therefore, x_n is equivalent to z_n.

Suppose x_n and y_n are p-adic Cauchy sequences. For every r, there are N and M such that

$$x_N \equiv x_{N+1} \equiv \cdots \bmod p^r \quad \text{and} \quad y_M \equiv x_{M+1} \equiv \cdots \bmod p^r.$$

If we let L be the bigger of N and M, then we have

$$x_L + y_L \equiv x_{L+1} + y_{L+1} \equiv \cdots \bmod p^r \text{ and } x_L \cdot y_L \equiv x_{L+1} \cdot y_{L+1} \equiv \cdots \bmod p^r.$$

Therefore, $(x_n + y_n)$ and $x_n y_n$ are p-adic Cauchy sequences.

Suppose x_n is equivalent to x'_n and y_n is equivalent to y'_n. If we fix an r, then for sufficient large n, we have $x_n \equiv x'_n \bmod p^r$ and $y_n \equiv y'_n \bmod p^r$. This implies $x_n + y_n \equiv x'_n + y'_n \bmod p^r$ and $x_n y_n \equiv x'_n y'_n \bmod p^r$. Therefore, $x_n + y_n$ is equivalent to $x'_n + y'_n$ and $x_n \cdot y_n$ is equivalent to $x'_n \cdot y'_n$.

4.38. There is an N such that for all $n, m \geq N$ we have $|x_n - x_m|_p \leq 1$. In particular, if $n \geq N$ then $|x_n - x_N| \leq 1$. This means that $x_n - x_N \in \mathbb{Z}_{(p)}$. Choose an $r \geq 0$ such that $p^r x_n \in \mathbb{Z}_{(p)}$ for all $n \leq N$. If $n > N$, then $p^r x_n = p^r x_N + p^r(x_n - x_N) \in \mathbb{Z}_{(p)}$.

4.39. Suppose that $\mathbb{Z}_p = \{\underline{x}_n : n \in \mathbb{N}\}$. Each of the p-adic integers has a unique expansion

$$\underline{x}_n = \sum_r a_{n,r} p^r, \quad a_{n,r} \in \{0, 1, \ldots, p-1\}.$$

To get a contradiction, we'll find a p-adic integer which is not in the list $\underline{x}_n$. To do this, we let $\underline{y} = \sum b_r p^r$, where the coefficients b_r are chosen so that $b_r \neq a_{r,r}$. By uniqueness of the expansion, this implies $\underline{y} \neq \underline{x}_n$.

5.1. In the first equation, y^6 is congruent to 0 or 1 modulo 7 by Fermat's Little Theorem, but $7x^3 + 2$ is congruent to 2 modulo 7. Therefore, there are no solutions.

In the second equation, any common factor of $31x + 30$ and $31x - 1$ would also be a factor of 31. However, $31x - 1$ is not a multiple of 31, so the two factors on the right-hand side are coprime. Therefore, $31x + 30$ and $31x - 1$ are both fifth powers, and they differ by 31. The pairs of fifth powers which differ by 31 are $1^5, 2^5$ and $(-2)^5, (-1)^5$. Therefore, $31x - 1$ is either -32 or 1. Since x is an integer, we must have $31x - 1 = -32$ so $x = -1$ and $y = 2$.

Factorizing the right-hand side of the last equation, we get $y^2 = x(x^2 + 1)$. The factors x and $x^2 + 1$ are coprime, so $x^2 + 1 = \pm d^2$ for some integer d. Clearly, the sign is $+$, so both x^2 and $x^2 + 1$ are squares. This implies $x = 0$, so $y = 0$.

5.2. The ring $\mathbb{C}[X]$ has unique factorization, so we may use Lemma 5.1. Since g and $g+1$ are coprime in $\mathbb{C}[X]$, we must have $g = cr^3$ and $g+1 = c's^3$ for polynomials r, s and constants $c, c' \in \mathbb{C}$. Since all complex numbers have cube roots, we can assume $c = c' = 1$. This implies $1 = s^3 - r^3 = (r-s)(r^2 + rs + s^2)$. It follows that $r - s$ is constant. If r has positive degree, then r and s have the same leading term. However, this implies that $r^2 + rs + s^2$ has positive degree, which gives a contradiction. Therefore, r is constant, so f and g are constant.

5.3. The ring $\mathbb{F}_3[X]$ has unique factorization so we may use Lemma 5.1. Note that every unit is a cube by Fermat's Little Theorem. The highest common factor of g and $g + X$ could be either 1 or X. If the highest common factor is X, then either g or $g + X$ is a multiple of X^2. In the first case, we have $r^3 = s(Xs + 1)$ with $f = Xr$ and $g = X^2 s$. This implies that both s and $Xs + 1$ are cubes. This can't happen when $s \neq 0$ because the degrees of s and $Xs + 1$ differ by 1, but it does happen if $s = 0$. This gives a solution $(f, g) = (0, 0)$. A similar argument shows that if $g + X$ is a multiple of X^2, then $(f, g) = (0, -X)$. In the other case, we are assuming g and $g + X$ are coprime, so both are cubes. This gives us a factorization $X = a^3 - b^3 = (a - b)^3$ where $g + X = a^3$ and $g = b^3$. This gives a contradiction since the degree of X is not a multiple of 3. Therefore, the only solutions are $f = 0$, $g = 0, -X$.

5.4. Think of $\mathbb{C}[X]$ as a subring of $\mathbb{C}[Y]$ where $X = -Y^2$. We have $f^3 = (g+Y)(g-Y)$. If $g+Y$ and $g-Y$ are coprime, then both must be cubes in $\mathbb{C}[Y]$. This gives a factorization $Y = a^3 - b^3 = (a-b)(a-\zeta_3 b)(a-\zeta_3^2 b)$. Hence, at least two of the three factors is constant, and so b must be constant, which easily gives a contradiction.

In the other case, $g + Y$ and $g - Y$ have a common factor Y, so in particular g is a multiple of X. This implies f is a multiple of X, and therefore $X = f^3 - g^2$ is a multiple of X^2. This also gives a contradiction, so there are no solutions.

5.5. Let $A = r + s\sqrt{d}$ and $B = x + y\sqrt{d}$. It's trivial that $\overline{A+B} = \overline{A} + \overline{B}$ and $\overline{\overline{A}} = A$. To show that $\overline{A \cdot B} = \overline{A} \cdot \overline{B}$, we expand out both sides of the equation:

$$AB = (r + s\sqrt{d})(x + y\sqrt{d}) = (rx + syd) + (ry + sx)\sqrt{d}$$

$$\overline{A} \cdot \overline{B} = (r - s\sqrt{d})(x - y\sqrt{d}) = (rx + syd) - (ry + sx)\sqrt{d}.$$

If $A = u + v\alpha \in \mathbb{Z}[\alpha]$, then u and v are integers. By what we've already proved, it follows that $\bar{A} = u + v\bar{\alpha}$. Since $\bar{\alpha}$ is in $\mathbb{Z}[\alpha]$, it follows that

$\bar{A} \in \mathbb{Z}[\alpha]$. Since $\overline{AB} = \overline{A} \cdot \overline{B}$, we have

$$\mathrm{N}(A \cdot B) = (A \cdot B)\overline{(A \cdot B)} = (A \cdot B)(\overline{A} \cdot \overline{B}) = \mathrm{N}(A) \cdot \mathrm{N}(B).$$

If $\mathrm{N}(A) = 0$, then $A\bar{A} = 0$. This implies $A = 0$ or $\bar{A} = 0$. However, if $\bar{A} = 0$ then $A = \bar{0} = 0$, so in either case $A = 0$.

5.6. If $A = r + s\sqrt{d}$, then $M = \begin{pmatrix} r & ds \\ s & r \end{pmatrix}$. Clearly, $\det(M) = r^2 - ds^2 = \mathrm{N}(A)$. We have $\det(XI_2 - M) = X^2 - 2rX + \mathrm{N}(A)$. If $A \in \mathbb{Z}[\alpha]$, then $2r$ and $\mathrm{N}(A)$ are in $\mathbb{Z}$. Conversely, suppose that $a = 2r$ and $r^2 - ds^2$ are integers. If follows that $4ds^2 = a^2 - 4\mathrm{N}(A)$ is an integer. Since d is square free, it follows that $b = 2s$ is an integer. We have $\mathrm{N}(A) = \frac{1}{4}(a^2 - db^2)$, and therefore

$$a^2 \equiv db^2 \bmod 4. \tag{5.12}$$

In the case that $d \not\equiv 1 \bmod 4$, Equation (5.12) implies that both a and b are even and so $A \in \mathbb{Z}[\sqrt{d}]$. In the case that $d \equiv 1 \bmod 4$, Equation (5.12) implies $a \equiv b \bmod 2$, and so $a = 2c + b$ for some integer c. This implies $A = c + b\frac{1+\sqrt{d}}{2} \in \mathbb{Z}[\frac{1+\sqrt{d}}{2}]$.

5.7. Let $A = r + s\alpha$ with $r, s \in \mathbb{Z}$. Then A is a root of either $m(X) = (X - r)^2 - ds^2$ or $m(X) = (X - r)^2 + (X - r)s + \frac{1-d}{4}s^2$ depending in d modulo 4. In either case, m is a monic polynomial with integer coefficients so A is an algebraic integer.

Conversely if A is a root of $X^2 + bX + c$, then $A = \frac{-b \pm \sqrt{b^2 - 4c}}{2}$. Let $b^2 - 4c = e^2 d$ with d square-free. If $d = 0$ or $d = 1$, then it's easy to show that $A \in \mathbb{Z}$. Assume $d \neq 0, 1$, so A is in the field $\mathbb{Q}(\sqrt{d})$. With this notation, we have $A = \frac{-b \pm e\sqrt{d}}{2}$. If b is even then e must also be even, and $A \in \mathbb{Z}[\sqrt{d}] \subseteq \mathbb{Z}[\alpha_d]$. If b is odd, then both d and e are odd, and $d \equiv 1 \bmod 4$. In this case, we have $A = \frac{-e-b}{2} + e\frac{1 \pm \sqrt{d}}{2} \in \mathbb{Z}[\frac{1+\sqrt{d}}{2}]$.

5.8. We find the quotients Q first:

$$\frac{18 + 5i}{7 + 3i} = \frac{141 - 19i}{58} \approx 2$$

$$\frac{33 + 7\sqrt{2}}{4 + 3\sqrt{2}} = \frac{-90 + 71\sqrt{2}}{2} \approx -45 + 35\sqrt{2}$$

$$\frac{40 + 6\alpha_5}{3 + 6\alpha_5} = \frac{-213 + 111\sqrt{5}}{9} \approx -36 + 25\alpha_5.$$

The remainders $R = A - QB$ are:

$$(18+5i) - 2(7+3i) = 4 - i,$$

$$(33+7\sqrt{2}) - (-45+35\sqrt{2})(4+3\sqrt{2}) = 2\sqrt{2}+3,$$

$$(40+6\alpha_5) - (-36+25\alpha_5)(3+6\alpha_5) = -2-3\alpha.$$

5.9. Let $Q = x + y\sqrt{-5}$. We have $R = A - QB = -2x + (1-2y)\sqrt{-5}$. Therefore, $\mathrm{N}(R) = 4x^2 + 5(1-2y)^2 \geq 5$, whereas $\mathrm{N}(B) = 4$.

5.10. Going through Eulcid's algorithm, we have

$$\begin{aligned} A &= (-1-2i)B + (3+i) \\ B &= (2+i)(3+i) + (-1+i) \\ (3+i) &= (-1-2i)(-1+i) + 0 \\ (-1+i) &= B - (2+i)(3+i) = B - (2+i)(A - (-1-2i)B) \\ &= (-2-i)A + (1-5i)B. \end{aligned}$$

5.11. We have $\mathrm{N}(Z) = x^2 - 6y^2 \in [-\frac{3}{2}, \frac{1}{4}]$. Assume $|\mathrm{N}(Z)| \geq 1$, so we have $\mathrm{N}(Z) \in [-\frac{3}{2}, -1]$. For $x \in [0, \frac{1}{2}]$, we have $\mathrm{N}(Z+1) = \mathrm{N}(Z) + 2x + 1 \in [2x - \frac{1}{2}, 2x]$, so $|\mathrm{N}(Z+1)| < 1$ unless $x = \frac{1}{2}$. In the case $x = \frac{1}{2}$, we have $\mathrm{N}(Z+1) \in [\frac{1}{2}, 1]$; this norm and cannot be 1 since y is rational. Therefore, $|\mathrm{N}(Z+1)| < 1$ in this case as well. Similarly, if $x \in [-\frac{1}{2}, 0]$ then $|\mathrm{N}(Z-1)| < 1$.

To show that the ring is norm-Euclidean, let $A, B \in \mathbb{Z}[\sqrt{6}]$ with $B \neq 0$. Choose Q_1 such that if $Z = \frac{A}{B} - Q_1 = x + y\sqrt{6}$ then $|x|, |y| < \frac{1}{2}$. We've shown already that $|\mathrm{N}(Z - Q_2)| < 1$ for some $Q_2 \in \{0, 1, -1\}$. Let $Q = Q_1 + Q_2$ and $R = A - QB$. Then we have $A = QB + R$ and $|\mathrm{N}(R)| = |\mathrm{N}(\frac{A}{B} - Q_1 - Q_2)| \cdot |\mathrm{N}(B)| < |\mathrm{N}(B)|$.

5.12. Suppose $x^3 = y^2 + 2$. This implies $x^3 = (y + \sqrt{-2})(y - \sqrt{-2})$. By the Disappointing Theorem, we know that $\mathbb{Z}[\sqrt{-2}]$ has unique factorization. If x is even then $y^2 \equiv -2 \bmod 8$, which gives a contradiction. Therefore, x is odd. Suppose Q is a factor of both $y + \sqrt{-2}$ and $y - \sqrt{-2}$ then $Q | 2\sqrt{-2}$, and therefore $Q|4$. Since x is odd, there exist integers h and k such that $4h + xk = 1$. Therefore, Q is a factor of 1, so Q is a unit. This shows that $y + \sqrt{-2}$ and $y - \sqrt{-2}$ are coprime. By Lemma 5.1, there is a unit $U \in \mathbb{Z}[\sqrt{-2}]$ and another element $a + b\sqrt{-2}$ such that $y + \sqrt{-2} = U(a + b\sqrt{-2})^3$. By Theorem 5.1, we have $U = \pm 1$; hence U is a cube, so we may assume that $U = 1$. Expanding out, we get $y + \sqrt{-2} = a^3 + 3a^2b\sqrt{-2} - 6ab^2 - 2b^3\sqrt{-2}$.

Equating the imaginary parts, we get $1 = 3a^2b - 2b^3$. In particular, b is a factor of 1, so $b = \pm 1$. This implies $3a^2 - 2 = \pm 1$, which has integer solutions $a = \pm 1$, $b = -1$. Equating the real parts, rather than the imaginary parts we get $y = \pm 5$, so the solutions are $(x, y) = (3, \pm 5)$.

The other equations are solved in the same way, and have the following solutions:

- $x^3 = y^2 + y + 2$. $(2, 2)$, $(2, -3)$.
- $x^5 = y^2 + y + 2$. $(2, 5)$, $(2, -6)$.
- $x^3 = y^2 + y + 3$. None.
- $x^5 = y^2 + y + 3$. $(3, 15)$, $(3, -16)$.
- $x^7 = y^2 + y + 5$. $(5, 279)$, $(5, -280)$.

5.13. The Minkowski bound is $M_{-19} = \frac{2}{\pi}\sqrt{19} < 3$, and $-19 \equiv 5 \bmod 8$. Therefore, S_{-19} contains no primes less than the Minkowski bound. By Theorem 5.4, $\mathbb{Z}[\alpha]$ has unique factorization. Only 19 is ramified. The primes $5, 7, 11, 17, 23$ split and the primes $2, 3, 13$ are inert.

$$
\begin{aligned}
5 &= 0^2 + 0 \cdot 1 + 5 \cdot 1^2 &&= \mathrm{N}(\alpha) &&= \alpha(1 - \alpha),\\
7 &= 1^2 + 1 \cdot 1 + 5 \cdot 1^2 &&= \mathrm{N}(1 + \alpha) &&= (1 + \alpha)(2 - \alpha),\\
11 &= 2^2 + 1 \cdot 2 + 5 \cdot 1^2 &&= \mathrm{N}(2 + \alpha) &&= (2 + \alpha)(3 - \alpha),\\
17 &= 3^2 + 1 \cdot 3 + 5 \cdot 1^1 &&= \mathrm{N}(3 + \alpha) &&= (3 + \alpha)(4 - \alpha),\\
19 &= 1^2 + (-2) \cdot 1 + 5 \cdot (-2)^2 &&= \mathrm{N}(1 - 2\alpha) &&= -(1 - 2\alpha)^2,\\
23 &= 1^2 + 1 \cdot 2 + 5 \cdot 2^2 &&= \mathrm{N}(1 + 2\alpha) &&= (1 + 2\alpha)(3 - 2\alpha).
\end{aligned}
$$

$1540 = 2^2 \cdot 5 \cdot 7 \cdot 11$. There are elements of norm 4, 5, 7 and 11, so there is an element with norm 1540. $399 = 3 \cdot 7 \cdot 19$. Tshe prime 3 is inert so there is no element of norm 399. $210 = 2 \cdot 3 \cdot 5 \cdot 7$. The primes 2 and 3 are inert so there are no elements of norm 210. $115 = 5 \cdot 23$. There are elements of norm 5 and 23, so there is an element with norm 115.

5.14. Let $\alpha = \frac{1+\sqrt{5}}{2}$. Only 5 is ramified in $\mathbb{Z}[\alpha]$. The primes which split are those which are congruent to 1 or 4 mod 5 (by the decomposition theorem and the Quadratic Reciprocity Law). We have $\mathrm{N}(x + y\alpha) = x^2 + xy - y^2$. Therefore, $-1 = \mathrm{N}(\alpha)$. It follows that n is a norm if and only if $-n$ is a norm. Therefore, it's sufficient to check whether $|\mathrm{N}(x + y\alpha)| = |n|$ has solutions. Solutions exist for in every case listed, because 5 is ramified and 41, 19 and 101 split.

5.15. Note that $\mathrm{N}(x + y\sqrt{3}) = x^2 - 3y^2$. If this is equal to -1, then $x^2 \equiv -1 \bmod 3$. However, this is impossible since -1 is not a quadratic residue modulo 3. Therefore, no element has norm -1. Suppose $\mathrm{N}(P) = \pm p$, so either p is ramified or p splits in $\mathbb{Z}[\sqrt{3}]$. In either case, the other elements of norm $\pm p$ are UP, $U\bar{P}$, where U is a unit. We've shown that no element has norm -1 so $\mathrm{N}(U) = 1$. Hence, $\mathrm{N}(UP) = \mathrm{N}(U\bar{P}) = \mathrm{N}(P)$. If $p \equiv 1 \bmod 12$, then by quadratic reciprocity, we have $\left(\frac{3}{p}\right) = 1$. Therefore, p splits, so there is an element with norm $\pm p$. Suppose $x^2 - 3y^2 = -p$. This implies $x^2 \equiv -1 \bmod 3$, which is impossible. It follows that there must be a solution to $x^2 - 3y^2 = p$.

Conversely, if $x^2 - 3y^2 = p$, then p is either ramified or split. Only 2 and 3 are ramified and we have $-2 = \mathrm{N}(1 + \sqrt{3})$, $-3 = \mathrm{N}(\sqrt{3})$. Hence, by the previous part of the question, $p \neq 2, 3$. Therefore, p is split. By the decomposition theorem, this implies $\left(\frac{3}{p}\right) = 1$. By the Quadratic Reciprocity Law, $p \equiv \pm 1 \bmod 12$. However, since $x^2 - 3y^2 = p$, it follows that $x^2 \equiv p \bmod 3$. Hence, $p \not\equiv -1 \bmod 3$ so we have $p \equiv 1 \bmod 12$. Note that $66 = 2 \times 3 \times 11$. The primes 2 and 3 are ramified and 11 is split. By the previous part of the question, there are element with norm $-2, -3, -11$ but not $2, 3, 11$. Therefore, solutions exist to $x^2 - 3y^2 = -66$, but not $x^2 - 3y^2 = +66$.

$$\begin{aligned} -66 &= (-2)(-3)(-11) = \mathrm{N}((1+\sqrt{3})\sqrt{3}(1+2\sqrt{3})) \\ &= \mathrm{N}(9 + 7\sqrt{3}) = 9^2 - 3 \times 7^2. \end{aligned}$$

Similarly, $1573 = 11^2 \times 13$ and there are elements with norm $-11, +13$. Therefore, solutions exist to $x^2 - 3y^2 = +1573$, but not $x^2 - 3y^2 = -1573$.

$$1573 = 11^2 \times 13 = 11^2(4^2 - 3 \times 1^2) = 44^2 - 3 \times 11^2.$$

5.16. We'll prove by induction on $|\mathrm{N}(A)|$ that A is either a unit or a product of irreducible elements. If $|\mathrm{N}(A)| = 1$, then we've already shown that A is a unit. Assume the result for elements with smaller norm than A. If A is irreducible, then there is nothing to prove, so assume $A = BC$, where neither B nor C is a unit. This implies that $|\mathrm{N}(B)|, |\mathrm{N}(C)| > 1$, and therefore $|\mathrm{N}(B)|, |\mathrm{N}(C)| < |\mathrm{N}(A)|$. By the inductive hypothesis, B and C are products of irreducible elements (they are not units because their norms are not 1). Therefore, A is a product of irreducible elements.

Suppose there are only finitely many irreducible elements $P_1, \ldots, P_r$ and let $N = |\mathrm{N}(P_1 \cdots P_r)| + 1$. None of the irreducible elements are factors of N, so by what we've already proved N must be a unit in $\mathbb{Z}[\alpha]$. This gives a contradiction since N is an integer and $N \geq 2$.

5.17. The rings $\mathbb{Z}[\alpha_{-63}]$ and $\mathbb{Z}[\alpha_{33}]$ do have unique factorization, but $\mathbb{Z}[\alpha_{34}]$ does not. For example, 3 should split in $\mathbb{Z}[\alpha_{34}]$ since $\left(\frac{34}{3}\right) = 1$. However, the norm of a general element is $x^2 - 34y^2$; this is never equal to ± 3 since $\left(\frac{\pm 3}{17}\right) = -1$.

5.18. Assume factorization is unique. We have $\left(\frac{pq}{l}\right) = 1$, so l splits. Therefore, there is an element of norm $\pm l$, so there exist integers or half-integers x, y such that $x^2 - pqy^2 = \pm l$. This implies $\left(\frac{\pm l}{p}\right) = 1$. By the First Nebensatz and the Quadratic Reciprocity Law, it follows that $\left(\frac{p}{l}\right) = 1$, which gives a contradiction.

5.19. Let $d = q_1 \cdots q_s$ be be the factorization of d. Then choose a satisfying the congruences

$$a \equiv 1 \bmod 4q_2 \cdots q_s, \quad a \equiv g \bmod q_1,$$

where g is a quadratic non-residue modulo q_1.

If p is a prime number congruent to a modulo $4d$, then by the Reciprocity Law,

$$\left(\frac{d}{p}\right) = \prod \left(\frac{q_i}{p}\right) = \prod \left(\frac{a}{q_i}\right) = -1.$$

Therefore, p is inert. Similarly if $p \equiv 1 \bmod 4d$, then p is split. By Dirichlet's primes in arithmetic progressions theorem, there are infinitely many such p.

5.20. $\frac{17}{45} = [0, 2, 1, 1, 1, 5]$, $\frac{103}{456} = [0, 4, 2, 2, 1, 14]$, $\frac{29}{135} = [0, 4, 1, 1, 1, 9]$.

5.21. In the case that the partial quotients are all 1, the recursive definitions of h_n and k_n are the same as the definition of the Fibonacci numbers. If we write $\alpha = \lim \frac{F_{n+1}}{F_n} = [1, 1, 1 \ldots]$, then we have $\alpha = [1, \alpha] = 1 + \frac{1}{\alpha}$. This implies $\alpha = \frac{1 \pm \sqrt{5}}{2}$. Since α is positive, we must have $\alpha = \frac{1+\sqrt{5}}{2}$.

5.22. We'll prove the continuity by induction on n. It's clearly true for $n = 0$. Assume it's true for n, so if we define $f : \mathbb{R}^{>0} \to \mathbb{R}$ by $f(x) = [a_0, \ldots, a_n, x]$, then f is continuous on $\mathbb{R}^{>0}$. Define $g(x) = a_{n+1} + \frac{1}{x}$. If $x > 0$, then g is continuous at x; furthermore $g(x) > 0$, so f is continuous at $g(x)$. It follows that $f \circ g$ is continuous at x. Since $f(g(x)) = [a_0, \ldots, a_{n+1}, x]$, this proves the result.

5.23. Let $\alpha = [2, 2, 2, \ldots]$. This implies $\alpha = 2 + \frac{1}{\alpha}$. Therefore, $\alpha^2 - 2\alpha - 1 = 0$, so $\alpha = \frac{1 \pm \sqrt{8}}{2}$. Since $\alpha > 0$, we have $\alpha = \frac{1}{2} + \sqrt{2}$. Similarly $2, 3, 2, 3, 2, 3, 2, 3, \ldots] = 1 + \sqrt{\frac{5}{3}}$ and $[1, 2, 3, 5, 3, 5, 3, 5, \ldots] = \frac{49+\sqrt{285}}{46}$.

5.24. We have $\alpha = [a_0, \ldots, a_r, \beta]$, where $\beta = [a_{r+1}, \ldots, a_{r+s}, \beta]$. Therefore, β satisfies an equation of the form $\beta = \frac{a\beta+b}{c\beta+d}$ with integers a, b, c, d. If we simplify this, we see that β is a root of a quadratic equation. Therefore, β is in one of the fields $\mathbb{Q}(\sqrt{\delta})$, where δ is an integer. From the way α is expressed in terms of β, it follows that α is also in this field, so we have $\alpha = r + s\sqrt{\delta}$ with rational coefficients r and s. From this, we see that α is a root of the polynomial $(X - r)^2 - s^2\delta$.

5.25. Let $Z = \frac{A}{B}$ with A and B coprime in $\mathbb{Z}[\alpha]$. We'll prove by induction on $|\mathrm{N}(B)|$. If B has norm ± 1 then B is a unit, so without loss of generality $B = 1$. In this case, we have $Z = A = [A]$. For the inductive step, divide A by B with remainder in $\mathbb{Z}[\alpha]$. We have $A = QB + R$ and $|\mathrm{N}(R)| < |\mathrm{N}(B)|$. By the inductive hypothesis, we have an expression of the form $\frac{B}{R} = [a_0, \ldots, a_n]$ with $a_i \in \mathbb{Z}[\alpha]$. Therefore, $Z = \frac{A}{B} = Q + (\frac{B}{R})^{-1} = [Q, a_0, \ldots, a_n]$.

5.26. The continued fraction expansions are $\sqrt{2} = [1, \overline{2}]$, $\sqrt{3} = [1, \overline{1, 2}]$, $\sqrt{5} = [2, \overline{4}]$, $\sqrt{6} = [2, \overline{2, 4}]$ and $\sqrt{10} = [3, \overline{6}]$. The fundamental solutions are $(3, 2)$, $(2, 1)$, $(9, 4)$, $(5, 2)$, $(19, 6)$. There is one obvious solution $(x, y) = (6, 1)$ to the equation $x^2 - 10y^2 = 26$. This implies $\mathrm{N}(6 + \sqrt{10}) = 26$. To get another solution, we multiply by an element with norm 1. Since $(19, 6)$ is a solution to Pell's equation, we have $\mathrm{N}(19 + 6\sqrt{10}) = 1$. Therefore, $(6 + \sqrt{10})(19 + 6\sqrt{10})$ has norm 26. Expanding this out, we get $\mathrm{N}(174 + 55\sqrt{10}) = 26$, so $(174, 55)$ is another solution.

Bibliography

[1] M. Agrawal, N. Kayal and N. Saxena, "PRIMES is in P", *Annals of Mathematics*, 160(2), pp. 781–793 (2004).

[2] J.-R. Argand, "Philosophie mathématique. Réflexions sur la nouvelle théorie des imaginaires, suivies d'une application à la démonstration d'un théorème d'analise", *Annales de Mathématiques pures et Appliqués*, tome 5, pp. 197–209 (1814–1815).

[3] A. Baker, *Transcendental Number Theory*, Cambridge University Press (1975).

[4] N. Childress, *Class Field Theory*, Springer-Verlag (2008).

[5] D. A. Cox, *Primes of the Form $x^2 + ny^2$: Fermat, Class Field Theory, and Complex Multiplication*, Wiley (2013).

[6] H. Davenport, *Multiplicative Number Theory*, Markham (1967).

[7] H. Davenport, *The Higher Arithmetic: An Introduction To the Theory of Numbers*, Hutchinson (1962).

[8] W. Diffie and M. E. Hellman, "New directions in cryptography", *IEEE Transactions on Information Theory*, 22(6), p. 644174.

[9] C. F. Gauss, "Demonstratio nova theorematis omnem functionem algebraicam rationalem integram unius variabilis in factores reales primi vel secundi gradus resolvi posse" (1799).

[10] C. F. Gauss, *Disquisitiones Arithmeticae* (1801).

[11] D. Goldfeld, "Gauss' class number problem for imaginary quadratic fields", *Bulletin of the AMS*, 13(1) (1985).

[12] J. A. Green, *Sets and Groups*, Chapman and Hall (1988).

[13] K. Ireland and M. Rosen, *A Classical Introduction to Modern Number Theory*, Springer-Verlag (1990).

[14] F. Lemmermeyer, *Reciprocity Laws: From Euler to Eisenstein*, Springer-Verlag (2000).

[15] Manin and Panchishkin, *Introduction to Modern Number Theory: Fundamental Problems, Ideas and Theories*, Springer-Verlag (2007).

[16] D. A. Marcus, *Number Fields*, Springer-Verlag (1995).

[17] S. J. Patterson, *Introduction to the Theory of the Riemann Zeta-Function*, Cambridge University Press (1995).

[18] C. Pomerance, "A tale of two sieves", *Notices of the AMS*, pp. 1473–1485 (1996).

[19] R. Rivest, A. Shamir and L. Adleman, "A method for obtaining digital signatures and public-key cryptosystems", *Communications of the ACM* 21(2), pp. 120–126 (1978).

[20] I. Stewart and D. Tall, *Complex Analysis*, Cambridge University Press (2008).

[21] I. Stewart and D. Tall, *Algebraic Number Theory and Fermat's Last Theorem*, Chapman and Hall (2015).

[22] A. Selberg, "An elementary proof of Dirichlet's theorem about primes in an arithmetic progression", *Annals of Mathematics* 50(2), (1949).

[23] W. Stein, *Elementary Number Theory: Primes, Congruences, and Secrets*, Springer-Verlag (2009) (free to download at http://wstein.org/ent/).

Index

S

T

U

V